D1601197

ADAPTING TO THE LAND

ADAPTING TO THE LAND

A HISTORY OF AGRICULTURE IN COLORADO

John F. Freeman
WITH
Mark E. Uchanski

UNIVERSITY PRESS OF COLORADO
Louisville

© 2022 by University Press of Colorado

Published by University Press of Colorado
245 Century Circle, Suite 202
Louisville, Colorado 80027

All rights reserved
Manufactured in the United States of America

The University Press of Colorado is a proud member of the Association of University Presses.

The University Press of Colorado is a cooperative publishing enterprise supported, in part, by Adams State University, Colorado State University, Fort Lewis College, Metropolitan State University of Denver, Regis University, University of Alaska, University of Colorado, University of Denver, University of Northern Colorado, University of Wyoming, Utah State University, and Western Colorado University.

∞ This paper meets the requirements of the ANSI/NISO Z39.48–1992 (Permanence of Paper)

ISBN: 978-1-64642-204-3 (hardcover)
ISBN: 978-1-64642-205-0 (ebook)
DOI: 10.5876/9781646422050

Library of Congress Cataloging-in-Publication Data

Names: Freeman, John F. (John Francis), 1940– author. | Uchanski, Mark E., 1980– author.
Title: Adapting to the land : a history of agriculture in Colorado / John F. Freeman, Mark E. Uchanski.
Description: Louisville : University Press of Colorado, [2021] | Includes bibliographical references and index.
Identifiers: LCCN 2021036657 (print) | LCCN 2021036658 (ebook) | ISBN 9781646422043 (hardcover) | ISBN 9781646422050 (ebook)
Subjects: LCSH: Agriculture—Colorado—History. | Agriculture—Environmental aspects—Colorado. | Agriculture—Economic aspects—Colorado—History. | Irrigation farming—Colorado—History. | Organic farming—Colorado. | Sustainable agriculture—Colorado. | Water rights—Colorado—History.
Classification: LCC S451.C6 F74 2021 (print) | LCC S451.C6 (ebook) | DDC 338.109788—dc23
LC record available at https://lccn.loc.gov/2021036657
LC ebook record available at https://lccn.loc.gov/2021036658

Front-cover photograph: Irrigating potatoes, 1900–1920. Courtesy, Denver Public Library, Western History Collection, MCC 1847.

We are unlikely to be the first species in the
world to be exempt from nature's limits.

—Richard D. Lamm, 2002

Contents

Acknowledgments

The principal source of historical material for this book is the Agricultural and Natural Resources Archive at Colorado State University. The collections are well curated with excellent finding aids, with much material digitized since our research was completed. We are most appreciative of the friendly assistance of Linda Meyer, Vicky Lopez-Terrill, and Clarissa Trapp. We also wish to recognize Dee M. Salo, interlibrary loan librarian at the University of Wyoming. The bibliography includes only a few of the extension circulars, short articles, and essays cited in the endnotes; those materials are readily available through the internet.

Because of the centrality of the experiment stations—renamed Agricultural Research, Development and Education Centers—and the Cooperative Extension Service to the history of Colorado agriculture, we express our deepest appreciation to CSU faculty and staff members for their insights: in particular, Mike Bartolo, Addy Elliott, Todd Hagenbuch, C. J. Mucklow, Frank Stonaker, and Steve Wallner. At the Colorado Department of Agriculture, Janis Kieft, Glenda Mostek, and Wendy White graciously responded to our many inquiries.

We especially want to thank Spike Ausmus, John Ellis, Matt Heimerich, Dan Hobbs, Jan and Virgil Kochis, and Bob and R. T. Sakata for taking time away from their farming and ranching to visit with us at length. Likewise, thanks to agriculturists Brad Erker (Colorado Wheat) and Andrew Hopp (Dupont Pioneer); to the late Bill Stevenson and former presidents Dave Carter and John Stencel of the farmers union; to Robert M. Skirvin, Mark's academic mentor at the University of Illinois; and to Ronald K. Hansen, Michael A. Massie, and Micah Richardson. Most important, we are deeply indebted to the anonymous reviewers for the University Press of Colorado and to the editorial and production staff members at the press for their sincere interest in the manuscript and their encouragement at every step during extraordinarily difficult times.

Finally, to help put agriculture and agriculturists within the context of the relationship between people and nature, we are most grateful for insights on agrarian interpretations of stewardship from Pastors Bryson Lillie, Stephanie Price, and Travis and Kristina Walker. From time to time in preparing a book or article, after reading a passage or participating in a single conversation, one comes across a proposition that strengthens, clarifies, and gives unity to one's own effort. Our deepest appreciation, therefore, goes to the late Colorado governor Richard D. Lamm for his willingness to engage with John Freeman in lengthy and far-reaching conversation.

ADAPTING TO THE LAND

Introduction

In 1883, pioneer publicist William E. Pabor prepared *Colorado as an Agricultural State: Its Farms, Fields, and Garden Lands*, which he intended as a practical guide for prospective farmers. He warned his readers that "those who reach Colorado with certain ideas of society, soil, climate, and country, based upon what they have left behind them, are likely to be disappointed." But once the intelligent, systematic farmer understands the methods for properly and judiciously cultivating the soil under irrigation, then nature would take care of the rest.[1] One would like to believe that Pabor, a friend of Horace Greeley and one of the founding members of the Union Colony in 1870, understood that the final measure of good farming was that one could farm again: in contemporary parlance, sustainable agriculture.

Since Pabor's book, much has been written about how to farm, raise livestock, grow fruit, and cultivate gardens in Colorado, but little has been written from a historical perspective. Alvin T. Steinel, an extension specialist at Colorado State University (CSU) in Fort Collins, did write a *History of Agriculture in Colorado* (1926) to commemorate the state's fiftieth anniversary.

https://doi.org/10.5876/9781646422050.c000

Robert G. Dunbar, while a history professor at CSU, contributed a fine essay on Colorado agriculture to Leroy Hafen's *Colorado and Its People* (1948). Later, for its cultural resources series on Colorado, the US Bureau of Land Management commissioned Frederic J. Athearn, Steven F. Mehls, and Paul M. O'Rourke to prepare four regional monographs (1980s). Each monograph contained substantial sections on agricultural history and mentioned the shift away from traditional agriculture using natural soil amendments to synthetic fertilizers, herbicides, and pesticides. Yet to come was a more deliberate shift meant to protect the ecological values that make those yields possible. The results of that latter shift may not be readily observable to the casual visitor over fields of commodity crops or miles of rangeland, but they are reflected in the burgeoning popularity of natural and organic foods and the resulting global organic food processing and distribution industry. The extent to which Colorado agriculturists adapted to or stretched beyond the limits of the land, all within the context of an increasingly urban society, is the subject of this book.

My own interest in agricultural history stems from my training in medieval and early modern European history, when the connection between villages and cities and their agricultural surroundings was far tighter than it is today. Since my real job made it impractical for me to spend significant amounts of time in Europe, I turned to relatively untouched archival collections close by. My first book concerned the civilizing role of horticulture, making life on the High Plains more livable. A peer reviewer suggested in passing that I prepare a complementary history of crop production on the High Plains. Because of my longtime fascination with Colorado, its varied topography, its climatic regions, and especially its demographic trends, I came to believe that an agricultural history of that single state would encompass continuities and changes throughout the region. Not only that, despite or perhaps because of what I see along the Front Range, I remain convinced that Colorado still has a chance to slow and even reverse seemingly unrestrained growth, accommodate nature's limits, and create a vibrant, earth-friendly society in which agriculture in all its aspects plays an increasingly significant part.

For a period of twenty years, I had the privilege of cultivating my own garden consisting of vegetables, small fruits, and a few orchard fruits. As in most of Colorado, my soils were grossly deficient in organic matter, which warranted continual experimentation with rotations, legumes, and green

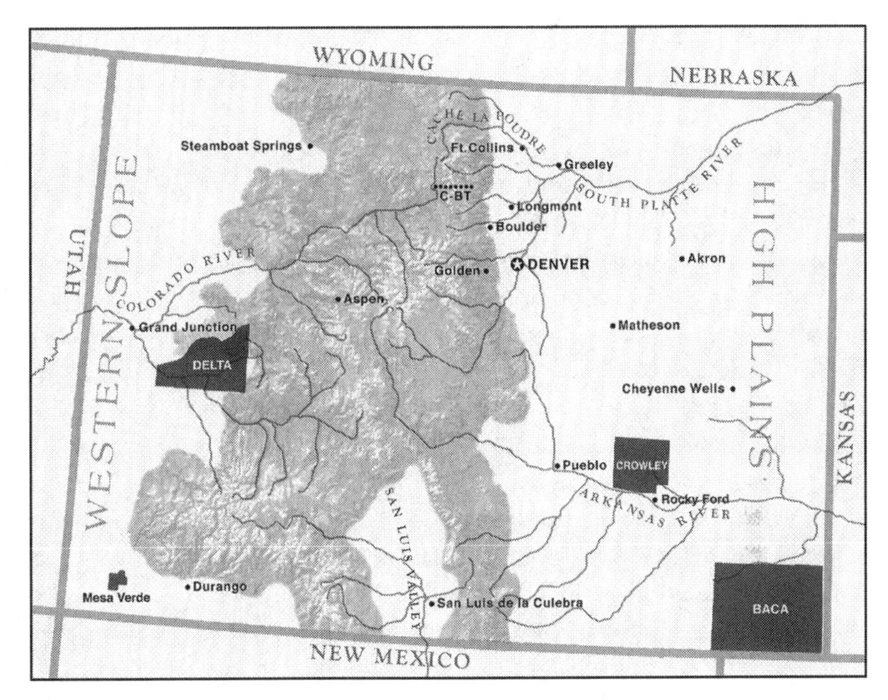

FIGURE 0.1. Colorado. *Courtesy*, Ronald K. Hansen.

manures. Playing at organic farming is recreational, good for mind, body, and soul. Farming to earn a modest living is quite another matter, as I discovered while preparing a history of the Rocky Mountain Farmers Union—longtime advocate on behalf of the family farmer, cooperative enterprise, and social democracy. In 2016 I attended the second annual High Plains Organic Farming Conference, co-sponsored by the agricultural extension services of Colorado State University and the University of Wyoming, where by chance I met Mark Uchanski, recently arrived from New Mexico State University to take the position of CSU's Specialty Crops Program coordinator. As Mark puts it, raised in Chicago's far western suburbs, where urban meets rural, he fell in love with farming and food systems, having earned his first few dollars picking strawberries on a nearby family farm, observing his grandfather's backyard garden, and tending to his own. He now directs a covey of graduate students, oversees his research at the CSU horticultural farm, and works cooperatively with specialty crop producers throughout the state.

He continues research that combines his interests in horticulture, ecology, and new and innovative approaches to maintainable cropping systems. I am especially grateful to Mark as an agricultural scientist for agreeing to help complete the present book.

Our story begins with Ancestral Puebloans, who had populated areas in today's southwestern Colorado. Archaeological evidence reveals that the Puebloans moved their settlements from time to time, perhaps to abandon deteriorated soils in favor of virgin soils. They did cultivate their crops under irrigation more than a millennium before Hispanic settlers introduced a system of communal irrigation to south-central Colorado. The 1859 gold rush created a market for food supplies that provided opportunity for farmers in the San Luis Valley to extend their primarily subsistence operations to commercial crops and attracted the first wave of farmers from the eastern states to establish market gardens along the Front Range and stock growers to move their livestock to the High Plains. During those early decades when soils were new to cultivation, farmers did not have to worry about conserving soil fertility; but as most Colorado soils tend to be alkaline, salty, clayey, or all three, they acknowledged the need to amend or "reclaim" their soils. Stock growers took advantage of free grazing for more than 1 million cattle and sheep spread over the vast rangelands that cover most of the state as well as the mountainous forestlands. By 1900, due in part to the advent of the railroads, Colorado agriculturists had established patterns of farming and livestock production (chapter 1).

As publicist, William Pabor featured those regions that could be irrigated by rivers and streams, pretty much ignoring grazing and forestry, which, in his opinion, had little to do with the economics of farming. Only with the administration of President Theodore Roosevelt did the term *agriculture* come to encompass grazing, forestry, gardening, and the culture of rural living as well as cultivating crops. While trying not to neglect any of those aspects, our book deals primarily with the cultivation of market produce and commodity crops.

Dry-land farming is perhaps the riskiest and most vulnerable method of cultivation when conducted without regard for the limits of land and water. With population growth nationwide, combined with the wartime (World War I) urgency to supply food overseas, federal policymakers pushed all farmers to vastly increase crop yields. The federal government financed basic research in the agricultural sciences, the application of that research through the experiment stations, and the dissemination of "useful knowledge" through the extension service—placing agents in counties to work directly with growers. Federal munificence positioned Colorado Agricultural College, the state's land-grant college, as the principal source of information on agriculture and related matters. As an active participant in the implementation of federal agricultural policy, the college generally responded to the needs of bigger farms and those efficiencies that produced higher yields and greater profits, although some voices cautioned against despoliation of land and water (chapter 2).

To retain high profits postwar, farmers borrowed heavily to purchase labor-saving machinery and expand land under cultivation. Turning to the exclusive use of pure seeds helped reduce weeds and increase yields, although certification contributed to reductions in the diversity of crops grown and to the cultivation of single-crop species to the exclusion of others. That, in turn, led to the increased use of chemicals against weeds and pests and to the introduction of manufactured inorganic fertilizers. Unprecedented drought exacerbated the downward spiral into the economic depression of the 1930s, a lesson yet to be fully learned on what happens when we seek to push nature beyond its limits (chapter 3). The impact of New Deal legislation on agricultural practices and rural living cannot be overstated. County agents were empowered to administer emergency relief and aid with temporary recovery. In the name of job creation and to protect against another economic disaster, the New Deal inaugurated broad reforms to conserve land and water. The Soil Conservation Act of 1935 provided the framework for soil restoration and other conservation activities—including the establishment of self-governing conservation districts to allay rural suspicion of federal intrusion—and for setting the foundation for future efforts to address the more pernicious aspects of industrialized agriculture. Taking advantage of federal funding for massive work projects, northern Colorado irrigators and municipalities lobbied successfully for the Colorado–Big Thompson Project (chapter 4).

World War II mobilization brought the "military-industrial complex" and an influx of population to Colorado. Jefferson County epitomized the transition from agricultural to urban development. The scientific and technical innovations that had been adopted to meet heightened wartime demand for food and fiber contributed postwar to further industrialization of agriculture: new plant cultivars and livestock breeds, advances in mechanization supported by computerization, and adoption of synthetic fertilizers, herbicides, and insecticides. The center-pivot system, invented by a Colorado farmer, greatly benefited crops under irrigation; combined with groundwater pumps, the system enabled the irrigation of dry lands from finite, previously untapped belowground resources. The scientific documentation and publicizing of the deleterious impact of toxic chemicals used in agriculture aroused public awareness of environmental despoliation generally, which led to the passage of landmark federal legislation. Widespread confidence expressed by most agriculturists that technology would allow people to overcome nature's limits was not universally shared. It kindled a conviction within academe and beyond that the resources of land, water, and all living systems are finite. Scientists and ethicists shed new light on stewardship of the "holy earth" and even encouraged a few farmers to challenge the economic, social, and ecological benefits of industrial agriculture (chapter 5).

On the public square, Colorado governor Dick Lamm espoused a broad understanding of stewardship as a balance between people and nature; he also advocated for planned rather than unfettered development and for conserving those values that make Colorado an exceptionally attractive place to live and work. An integral part of his plan for limiting economic development was the preservation of farmland and ranchland and making it more difficult to transfer irrigation water for municipal purposes. Starting in the late 1970s, the influx of affluent, well-educated professionals to the Front Range and the mountain towns contributed to passage of county-wide land-use regulations, supportive taxes, and increased use of tools such as conservation easements. Despite notable local efforts to bring about a balance of people and nature, continued growth in population and economic development compelled cities to redouble their efforts to secure more water and incentivized agriculturalists to sell their land and accompanying water rights to municipalities (chapter 6).

Changing consumer preferences nationwide moved the US Congress to begin supporting research, instruction, and extension activities in what became known as low-input sustainable agriculture—its purest form yet known being organic agriculture. Not coincidentally, the influx of urbanites to Colorado helped create a market niche for natural and organic foods locally grown. That, in turn, helped spur incremental changes in conventional farming; drew younger, often inexperienced idealists to experiment with intensive, small-scale farming; and attracted socially minded entrepreneurs. Boulder is considered the county of origin for the nation's organic and natural food distribution systems. Controversy over bioengineered crops placed Boulder at the epicenter of debate over ecologically acceptable agricultural practices and, more generally, over what makes for a healthy, vibrant, and enduring community and, by extension, a more livable world (chapter 7).

At the outset, a word about the use of the terms *organic farming* and *sustainable agriculture*. Prior to the twentieth century, nearly all farming was organic by necessity. As organically grown foods became a consumer choice, organic farmers wanted to differentiate their products from natural and conventionally grown foods and to assure consumers that their products started with organic seed or organic transplants and were grown without synthetic amendments and without antibiotics and growth hormones in the case of animal products. The 1989 Colorado legislature authorized an organic foods certification program; this program was later adjusted to comply with rules made as a result of federal legislation and is brought up to date on a regular basis.

Sustainable agriculture, alas, has become one of those fashionable catchwords with little precise meaning, with no federally approved certification program. The National Academy of Science, however, has defined the term to mean "an integrated system of plant and animal production practices" that over the long term satisfies human food and fiber needs, enhances environmental quality, makes efficient use of nonrenewable resources, sustains the economic viability of agriculture, and improves the quality of life for agriculturists and society in general.

To be sure, at least since ancient Rome, agricultural writers have expressed some understanding of farming and grazing to meet current needs without diminishing the ability of future generations to meet their needs. In Colorado beginning with the Greeley colonists, outspoken farmers have advocated for agriculture that is enduring, economically profitable, and socially responsible. The closer we get to the present, the greater the emphasis on what agricultural scientists would call regenerative agriculture—preserving and improving soils and waters; in sum, the overall ecology—although economic and social factors remain essential. Sustainable agriculture has become a goal as well as a system.

1

Agricultural Foundations

On a Friday evening in December 1870, about fifty members of the Union Colony met in the colony hall to organize the first Farmers' Club in the Territory of Colorado. Members debated whether to unite with the lyceum, become a branch of the Colorado Agricultural Society, or organize as an independent group. According to the *Greeley Tribune*, members voiced strong opposition to uniting with the lyceum because "men simply gifted as speakers, and with literary qualifications, but who had never raised a bushel of potatoes, and who had no real living interest in agricultural pursuits . . . would be forever talking about matters they do not understand." Members voted unanimously for an independent club, fixed membership dues at one dollar, and elected Captain David Boyd as president.[1]

A native of Ireland, Boyd had immigrated to Michigan with his family in 1851. He enrolled in the classical course of studies at the University of Michigan, left to serve in the Union Army, and returned after the Civil War to complete his degree. After college, he got married and bought a farm in Michigan, which he worked and then sold to move in March 1870 to the

https://doi.org/10.5876/9781646422050.c001

Union Colony—not a colony in the modern sense but a new town in the ancient Roman use of the term. Among the earliest members of the colony, Captain Boyd would succeed Nathan Meeker as president of the colony board of trustees. Boyd's prominence stemmed from the elected offices he held within the colony and later as state legislator; his principal legacy was his contribution to the management of water. Three years after Colorado statehood, which occurred in 1876, he presided over the select committee that drafted the procedures for the regulation of state water rights. Farming on semiarid land miles far from the foothills, Captain Boyd and his fellow colonists had learned early on that their claims on Cache le Poudre River water would remain uncertain until all claims to that water were defined, made a matter of written record, and then enforced by the appropriate authorities. Water being the lifeblood of agriculture, no community of farmers would contribute more to the early development of Colorado agriculture than the Union Colony.[2]

As a keen observer, Captain Boyd also understood the limits of irrigation. He criticized land agents who told prospective farmers that they should expect to harvest 26 to 28 bushels of wheat per irrigated acre and up to 70 bushels with special care when colony farmers considered a good harvest to be 20 bushels per acre. He disputed the common notion that irrigated land never loses its fertility. He argued instead that irrigation water does not deliver enough of the basic nutrients to grow crops and restore the soil after harvests, so "continual cropping, without returning anything to the soil, gradually exhausts it." As a result, the farmer's task is to assist nature by putting back into the soil those lost ingredients that contain the nutrients required to maintain soil fertility. Captain Boyd firmly believed in sustainable agriculture long before that term became popular.[3]

For the early inhabitants of the area known today as Colorado, cultivated plants and domesticated animals served only as supplement to their diets of wild plants and animals. Indeed, it was unlikely that any cultivation took place on the vast grasslands. We would like to imagine that the major source of food for indigenous plains dwellers was bison and, to a lesser extent, other large ungulates such as elk, deer, and antelope. But archaeologists and paleobotanists have discovered that wild plants gathered seasonally were the staff of life. In a useful guide for present-day gatherers, ethnobotanist Kelly Kindscher documented 123 species of native plants used as food by ancient plains peoples. Among the more important flowering plants: lamb's quarter (*Chenopodium berlandieri* Moq.), gathered for tender spring greens in spring and seeds in the fall; bush morning glory (*Ipomoea leptophylla* Torr.), for large roots similar to the sweet potato; prairie parsley (*Lomatium foeniculaceum* [Nutt.] Coult & Rose), for its starchy roots used to make flour; and the prairie turnip (*Psoralea esculenta* Pursh.), for its nutritious roots that could be preserved through winter. Botanists have documented hundreds of other edible species including familiar woody plants such as the wild plum (*Prunus americana* Marsh.), chokecherry (*Prunus virginiana* L.), and serviceberry (*Amelanchier alnifolia* Nutt.).[4]

For the very beginnings of agriculture in the region, we need to look to southwestern Colorado, where archaeologists have unearthed evidence of farm settlements dating back to 600 CE. The earliest settlements were apparently located in the Dolores River Valley, where the Ancestral Puebloans, or Eastern Anasazi tribes, farmed on flood plains that restored themselves through annual flooding. By the time of the famous cliff dwellings, preserved at Mesa Verde National Park, population growth had compelled farmers to expand their cultivation to terrace-like plots at the heads of canyons and to the tops of mesas where the soils were less fertile and more subject to drought. By then, however, farmers had learned to cope with the scarcity of water by constructing catchments and ditches for both domestic and agricultural use.[5]

These early farmers planted three principal crops: corn (*Zea mays* L.), beans (perhaps *Phaseolus acutifolius* A. Gray), and after about 900 CE, several types of squashes (*Cucurbita pepo* L.). Corn stalks served as trellises for the bean plants to climb; squashes helped preserve moisture by covering the soil with large leaves. We now know that corn depletes soil fertility while beans regenerate the soil, that the amino acids in beans help us digest the proteins in corn. The Three Sisters of corn, beans, and squashes not only made for a nutritionally balanced diet but suggested the extent to which these original stewards of the land adapted to it. By 1300, after a long period of drought, the Ancestral Puebloans apparently migrated south into the Chaco Canyon area of present-day New Mexico.[6]

In recent years, growers and seed companies have successfully marketed as an heirloom type the dry bean variety thought to have been grown by the Ancestral Puebloans. Kidney shaped, this bean is colored a striking

FIGURE 1.1. Cliff Palace, Mesa Verde, 1925. *Courtesy*, Denver Public Library, Western History Collection, MCC 3060.

dark red or burgundy with creamy-white blotches. High in protein, it has a mild, sweet flavor, requires no soaking, and cooks more quickly than its far more common cousin, the pinto. The fact that it is marketed under several names—Anasazi, Aztec, New Mexico Cave, New Mexico Appaloosa, Jacob's Cattle—makes one wonder about the bean's true identity, all of which raises the issue of nomenclature. For plants created in nature, botanists have established a universally recognized system for classification—identifying plants according to their relationship to each other and the names of authors who first described them. Hierarchically from the most general to the most specific, botanists classify plants at six levels; the last three are family, genus, and specific epithet (species). Because Latin was the universal language of scientists, they recognized the family name of a plant by the ending *aceae*; for example, the Anasazi bean belongs to the family of legumes known as Fabaceae. Within that family, the Anasazi bean belongs to the genus *Phaseolus*, meaning bean with an edible pod, the genus name always given as a Latin noun. Within that genus, the specific name is written as a Latin adjective, in this case *vulgaris*, meaning common, of frequent occurrence. It turns out that the eighteenth-century Swedish botanist Carl Linnaeus (abbreviated L.) first described the common bean with an edible pod as a separate species. Thus the full name of that bean became *Phaseolus vulgaris* L.

Botanical classification takes us only so far. In addition to plants created in nature, agriculturists have developed a vast number of varieties through plant propagation and plant breeding. We may well imagine that many millennia ago, farmers began to domesticate *Phaseolus vulgaris*, resulting in a number of cultivated varieties. The names of these cultivars are generally given in English, capitalized, and written in single quotes, such as *Phaseolus vulgaris* L. cv. 'Anasazi'—except that the Anasazi bean apparently is not broadly enough known to have earned an exclusive cultivar name. To further complicate matters, the classification of plants remains fluid, changing as a result of new research. Worse yet, for crop varieties developed through cultivation, the use of common names is governed by no formal code and may vary from region to region, seed house to seed house, nursery to nursery.

A truism often worth repeating is that farming requires permanent settlement, that is, some form of community. After the Ancestral Pueblo culture vanished into obscurity, various waves of hunters and gatherers occupied its area, leaving no evidence of organized farming. Spanish explorers encountered nomadic peoples as early as the mid-seventeenth century, but Hispanic farmers did not settle permanently in present-day south-central Colorado until the mid-nineteenth century. By 1851, twenty-five years before Colorado statehood, Hispanic farmers from the Taos Valley had established the first permanent agricultural settlements along the Culebra River, a tributary of the Rio Grande, in the San Luis Valley.

These farmers brought with them seeds of domesticated plants indigenous to the Americas (beans, corn, chile peppers, and potatoes) and seeds long ago introduced from Europe (lentils, peas, and wheat). Their corn became known as Mexican corn to differentiate it from corn cultivars imported by settlers from the eastern states. Their wheat was what the English called corn, although the French had used the same name (*blé*) for corn and wheat. Linnaeus had classified the former as *Zea mays* and the latter as *Triticum aestivum*, both belonging to the family of grasses (Poaceae). The Hispanic farmers also brought in livestock: beef and dairy cattle, sheep, goats, hogs, chickens, horses, mules, donkeys, and oxen. By importing such a diversity of livestock, farmers could provide for transport and supplement their own muscle power with animal power as well as fulfill their food and fiber needs.[7]

Just as significant for agriculture as was importing plants and livestock, Hispanic farmers introduced the institutions of communal governance and mutual aid. The village of San Luis de la Culebra, recognized as the oldest continuously occupied municipality in Colorado, was platted in the Spanish style with private dwellings situated around a central square, leaving room for churches and communal buildings. Some farmers walked daily from the village to their farms and home again at the end of the workday; others lived on the land they cultivated. Just who owned the land and what part belonged to the commune remained obscure until after statehood. Likewise, the management of water differed substantially from assigning water to landowners along waterways, as occurred in the more humid eastern states, and what would prevail in Colorado after statehood. Under the *acequia* system, the municipality managed the distribution of water; each user obtained an equitable share. Farms in the San Luis Valley were located along streams or ditches, with a succession of gardens, fields, and common pastures stretching in increasingly broad strips away from those waterways.[8]

Using water from the Culebra River, the San Luis People's Ditch served as the primary ditch (*acequia madre*) for the village of San Luis and surrounding farms. Originally dug by hand, the ditch was later deepened, widened, and extended using oxen pulling plows. Farming was viewed as a community endeavor, so village authorities (*alcaldes*) granted farmers specific times and allotments to fill their secondary ditches and, from there, to direct water through temporary channels into their gardens and fields. Water allocations, construction, and maintenance of the primary ditch depended on cooperation by all water users. The San Luis People's Ditch holds the first recorded water right in Colorado, dated April 10, 1852, and confirmed in 1889.

For farming, the principal tool used by Hispanic settlers in the San Luis Valley was the hoe. Only occasionally did they use a primitive plow, which was made entirely of wood, including the share or cutting part: a plow did not last long. Grains were sown by hand, harvested by sickle or infrequently by scythes, and threshed with the help of goats that trod out the grains on makeshift threshing floors. At first, women did the milling, placing the edible parts of grains on metates (flat or hollowed stones), on which they ground the grain using smaller stones. By the mid-1850s, the first Mexican mills, so-called in contrast to American mills that produced more refined flour, operated using water wheels in the southern San Luis Valley.[9]

Farmers in the San Luis Valley knew that integrating livestock into farm operations contributed to soil fertility, just as the ancients had understood. Three millennia ago Homer sang about Odysseus's dog Argos, which, in the absence of his master, lay on a deep pile of mule and oxen dung, ready for Odysseus's servants to spread manure over the fields of his estate (*Odyssey* 17: 290–300). During the first century CE, the Roman agriculturist Columella described the use of manure as nourishment for the soil: on such food, he wrote, the land grows fat (*On Agriculture* 2: 5).

With the advent of the Pikes Peak gold rush in 1859, farmers in the San Luis Valley began supplying miners with flour, which according to Horace Greeley was one of the four necessities in the execrable miner's diet in addition to whiskey, coffee, and bacon. The high cost of importing foodstuffs over vast distances in the pre-railroad era provided an incentive for unsuccessful miners as well as farmers looking for virgin soils to put down roots. Isolated attempts at farming had taken place along rivers on the eastern plains during the 1820s and 1830s—for example, at Bent's Fort along the Arkansas River and Lupton's Fort on the South Platte River. Such early efforts at farming failed due to the transient nature of the outposts and to the continuing presence of nomadic tribes.[10]

The gold rush contributed to the first permanent settlements along the Front Range, the narrow area on the western edge of the High Plains bordering the foothills of the Rocky Mountains; today, the Front Range is the most densely populated area of Colorado, extending roughly from the Wyoming state line on the north to the City of Pueblo on the south. Let us not forget that when easterners first began moving to Colorado, the issue of land ownership was murky at best. According to accepted European law, the US government had gained sovereignty as a result of the Louisiana Purchase in 1803, further clarified for southern Colorado by treaty with Mexico in 1846—from which it followed that the federal government possessed the exclusive right to purchase land from the tribes, recognized as the land's lawful owners. In principle, then, the early setters along the Front Range were squatters. To obtain some security in the absence of functioning political institutions, these agriculturists did what Americans have always done to solve big problems: come together in voluntary associations, in this case the mutual protective association known as the claim club, a specific application of the principles underlying the nation's founding documents. The preamble to the constitution of the Arapahoe County Claim Club (est. 1859) read: "Whereas it sometimes becomes necessary for persons to associate themselves together for certain purposes, such as the protection of life and property; and as we have left the peaceful shade of civilization—left friends and homes for the purpose of bettering our condition, we, therefore, associate ourselves together." Whereupon members stated their claims, gave geographic descriptions, and, to confirm the validity of claims, agreed to "make, or cause to be made, improvements on his or their claim, by breaking one acre of land; or building a house sufficiently good to live in." By 1861 claim clubs covered most of the Front Range and had gained recognition by the US Congress and the territorial legislature, Colorado Territory having just been carved out of portions of four existing territories. By resolving conflicts over land until state and federal laws took hold, the claim clubs contributed to the permanence of commercial agriculture as well as settlements.[11]

David K. Wall is considered the first commercial vegetable grower to use irrigation in Colorado. He arrived from Indiana by way of California, establishing a market garden on 2 acres along Clear Creek near Golden. In a May 1859 issue of the *Rocky Mountain News*, William N. Byers placed a notice that Wall had dropped off a large supply of vegetable seeds, which Byers packaged and advertised for sale. Three months later Byers reported that Denver "is now well supplied with garden vegetables of as fine quality as can be found in the old settlements of the States."[12] Byers himself had recently arrived from Omaha, Nebraska, where he had worked as a surveyor, bringing with him by oxcart a small used printing press with which he started Colorado's first newspaper. Soon thereafter, he laid claim to 160 acres along the South Platte 2 miles from Denver in what became known as the Valverde neighborhood. In 1862 his foreman had 50 acres under cultivation, yielding "melons of every variety, vegetables of all kinds—potatoes, tomatoes, cabbages, onions, egg plants, beets, peas, beans and everything that grows in this country." The farm produced enough, according to a visitor, so that if the newspaper business failed to provide sufficient income, Byers could still enjoy "the good things of life."[13]

A bit further distant from Denver, 4 miles north along Clear Creek, H. H. McAfee of B. McCleery and Company provided the *Rocky Mountain News* with the first report on raising commodity crops, which were field crops grown in relatively large amounts and sold to processors rather than directly to wholesalers and consumers. Under his direction, farm laborers had planted 46 acres of wheat, rye, barley, corn, oats, millet, sorghum, and potatoes. They dug a small ditch from the creek to the wheat field, but carters driving wagons inadvertently broke down the banks, causing the water to be deflected and this particular irrigation project to be abandoned. Nonetheless, the wheat matured with the "best filled [heads] seen in any country."[14] McAfee took samples to the *Rocky Mountain News* office, explaining that his crops had been grown without irrigation because sufficient rain had fallen during the growing season.[15]

The *Rocky Mountain News* served as the principal chronicle of early farming in Colorado. Beyond reporting, publisher Byers editorialized for agriculture as the most promising sector of the region's economy. He joined other Denver civic leaders in forming the Colorado Agricultural Society in March 1863 and was elected its secretary. The group's declared purpose was the advancement of farming and livestock production; its true purpose, however, was the promotion of a territorial fair to exhibit Colorado products. As a business development organization, the society attracted little farmer support, perhaps a harbinger of the differences between those who view farming as a business meant to generate maximum profits and others who consider farming to provide social and cultural as well as economic benefits.[16]

Commercial orchards were more difficult to establish than market gardens or fields of commodity crops. The costs and risks of importing nursery stock by wagon over long distances were higher; snow and late frosts proved disastrous to some stock. Nonetheless, by the early 1870s, orchards flourished all along the Front Range. Their success resulted from the diligence of orchardists such as Joseph Wolff of Boulder County; he had imported root stock by wagon from points east. During the harsh winter and spring of 1872–73, he had lost his entire orchard of 300-plus fruit trees to cold, dry winds. Yet he was not discouraged. "Fruit culture in Colorado," he wrote, "is a system of experimenting, and must for many years be largely in that condition, until experience shall determine what varieties to plant, the soil required, and proper tillage, the effect of irrigation, mulching, fertilizers, and other equally as important matters." No better statement of justification could be given for establishing the future agricultural experiment station at Fort Collins.[17]

The farmers from the eastern states who followed the "Fifty-Niners" had skipped over that vast expanse of the Great Plains once called the Great American Desert, from roughly the 100th meridian running through Nebraska and Kansas on the east to the Rocky Mountain foothills on the west. But earlier, explorers had reported on the groundcover that sustained the abundance of native ungulates—bison, elk, deer, and antelope—and provided the forage required by the domesticated ungulates—cattle, sheep, horses, oxen—those explorers had brought with them. "With occasional exceptions," John C. Frémont wrote, "these prairies are everywhere covered with a close and vigorous growth of a great variety of grasses, among which the most abundant is the buffalo grass."[18] Indeed, buffalo grass (*Buchloë dactyloides* [Nutt.] Engelm.) is the dominant short grass of the High Plains, growing less than 8 inches tall, while blue grama grass (*Bouteloua gracilis* [H.B.K.] Lag. ex Griffiths) dominates the driest prairie land, growing up to 2 feet. Both grasses are perennials, enduring year after year. The ongoing attempt by geneticists at the Land Institute in Salina, Kansas, to breed perennial grains for human consumption and thereby contribute to preserving soil fertility is among the most exciting and farsighted ongoing agricultural experiments on the entire Great Plains.

To supply miners and settlers with meat, stock growers trailed cattle up the Arkansas River Valley, along Fountain Creek, and over the Palmer Divide to Denver; in addition, porters, or freighters as they were called in the West, sold their worn-out oxen for meat. A former miner named Samuel Hartsell started a business of finishing cattle for market by fattening oxen on the grasses of his South Park ranch and by importing from Iowa the first herd of purebred cattle—shorthorns that could be raised as either beef cattle or dairy cows. The first sales of fattened cattle on the Denver market occurred in 1862. Settlers from the East, meanwhile, had trailed their dairy cows to the Front Range, so milk, butter, and cheese were available from the beginning; dairy-based businesses such as cheese factories did not appear until the 1870s.[19]

After the end of the Civil War, stock growers began trailing cattle from Texas to Colorado Territory's eastern plains. On that expanse of un-appropriated federal land, cattle could graze year-round because, unlike grasses in humid regions that break down with frost, western grasses remain erect and dry,

FIGURE 1.2. Livestock on land grant, San Luis Valley, 1900–1925. *Courtesy*, Denver Public Library, Western History Collection, MCC 4153.

providing nutritious forage during winter months. Cattle roamed unfettered in the days before fencing, requiring no attention until gathered together for shipment to markets. The first rail transport of cattle out of Colorado took place in 1869 from Kit Carson, at that time the western terminal of the Kansas Pacific Railway.[20] With the introduction of barbed wire in the 1870s, large cattle operators took advantage of loopholes in the Homestead Act of 1862 by filing claims along streams and water holes—often in the names of their hired men—thereby obtaining de facto control of vast hinterlands and preventing any type of farming. The harsh winter of 1885–86 destroyed about half of the range herds; that combined with President Grover Cleveland's executive order to remove all fences on public lands caused most large cattle operations to go out of business.[21]

During the late 1860s, the railroads had begun to crisscross northern Colorado, enabling market gardeners to ship their produce from places such as Golden, Boulder, and Loveland to the Denver market. It was not until 1870, however, that the Kansas Pacific connected Denver and southeast Colorado to markets in Kansas City and points east and the Denver Pacific connected Denver to Cheyenne, Wyoming, and the transcontinental rail line. To spur development, the federal government had granted the railroads rights-of-way many miles wide. In the late fall of 1869, the Union Colony of Colorado, a corporation created in New York, had taken an option to buy roughly 50,000 acres, located 60 miles north of Denver and owned by the real estate subsidiary of the Denver Pacific. Thus began a most remarkable social and economic experiment to create an ideal agricultural community. Although it failed to live up to the expectations of its founders, Greeley would become the seat of Colorado's most prosperous agricultural county.

Horace Greeley, the founder, editor, and publisher of the *New York Tribune*, may not have come up with the idea of the Union Colony but he publicized it, believing that migrating to the West offered impoverished farmers, laborers, and others a chance to start anew as members of a community neither too small nor too large. There, through cooperation and education, residents could meet all their economic, social, and cultural needs and live in harmony and happiness. The leader chosen for this noble venture was the quixotic Nathan C. Meeker, farmer and journalist, whose agrarian zeal ultimately cost him his life. It is not too much of a stretch to suggest that from time to time his idealistic spirit would resurface, whether through progressive farm associations, religious denominations, or, most recently, the movement for natural and organic foods locally grown.

The Union Colony location committee laid out the Greeley town plat along the Cache le Poudre River, approximately 1 square mile (640 acres) consisting of business and residential lots and reserving some land for the town square, schools, churches, and other public institutions. Town lots were "so laid out to be sold, one of each to each member of the Colony, at a fixed valuation, and the proceeds devoted to improvements for the common welfare." In addition, tracts adjoining the town site were "to be divided into lots of five, ten, twenty, forty and eighty acres, according to their distance from the town center, and deeded one to each member as they may choose or the [colony's executive] committee may decide." Upon the initial distribution of land, each member became a property owner on an equal basis with the others.[22]

Since low-lying or bottomland along the Cache la Poudre River had been taken by pre-colony settlers and most of the colony's acreage was benchland, it quickly became apparent that importing water through canals and ditches would be essential to the colony's farming success. Recognizing that water determined the value of land, the executive committee, later known as the Greeley board of trustees, adopted the principle of tying water rights to land title; in its communitarian spirit, the committee decreed that canals and ditches providing water to the colony be built, owned, and managed by the colony, not by individual water users. The first canal, completed in 1870 and known as Greeley #3, was a short-distance diversion from the south side of the Cache la Poudre onto town lots and gardens. On the north side, Greeley #2 took out water several miles above the town and ran for 36 miles. Although some irrigation on bottomlands had occurred in Colorado before 1870, Greeley #2 is considered the first canal to have provided irrigation water to benchland and Greeley the first community to have developed a comprehensive system of irrigation.[23]

The original plat of Greeley reflected Nathan Meeker's dream of a self-sufficient agricultural community surrounded by small-scale farms producing a variety of vegetables, small fruits, and nursery stock. In 1875 Meeker expressed the opinion that Colorado would never more than double its existing 100,000 acres of irrigated land but that such doubling would produce an immense amount of grains and vegetables. He reminded his readers that "most of the fruits and vegetables that supply the city of Paris, with its population of nearly three millions, come from only 3,000 acres. These acres, however, are in small parcels, broken up less by the plow than the spade, and no other land in the world yields so enormously, unless it be the irrigated gardens of northern Italy."[24]

Two-thirds of the original union colonists, according to David Boyd, intended to go into intensive farming on a few acres, but Bryant S. LaGrange was not among them. He believed colony farmers, who were close to Denver, would find growing commodity crops such as wheat and potatoes more profitable than garden produce. At the time colony plots first went on sale, the executive committee denied La Grange's request to purchase 160 acres. Instead, he was allowed a maximum of 80 acres, though in time he accumulated many more. A chatty newspaper columnist reported that by 1877, LaGrange was cultivating more than 200 acres: 150 acres in wheat, 35 in oats,

FIGURE 1.3. Irrigation, 1905. *Courtesy*, Denver Public Library, Western History Collection, MCC 1255.

25 in corn and potatoes, and the rest in "garden stuff." Six miles of ditches and laterals fed water to his fields and piped running water to his home and dairy cellar. Recognized as the first major producer of commodity crops under irrigation in Colorado, LaGrange was later appointed by Governor John L. Routt (R) to the first State Board of Agriculture. While LaGrange did well as a commodity producer, Nathan Meeker failed to make a modest living from a market garden and nursery stock on his 5 acres. His dream of a self-reliant community surrounded by small diversified farms never came true; Greeley would develop as the center of commodity and livestock production in Colorado.[25]

As the irrigated benchlands became consolidated into larger and larger fields, the humble potato cultivar known as the 'Greeley Spud' emerged as the principal commodity crop. The cultivar was grown in rotation with the legume alfalfa (*Medicago sativa* L.), a newly rediscovered soil restorative but dependent on irrigation, which in time became a major ingredient in the finishing ration for livestock. Even before the colonists arrived in Greeley, early settlers had brought with them seed potatoes, varieties unknown, that they had planted in their domestic gardens. At certain times in the past, the potato has been virtually the only source of food for human consumption. Although it can be grown under a variety of soil and climate conditions, its cultivation poses many difficulties, some continuing to this day. Captain Boyd observed that for the first two years of the Union Colony, potatoes grew well in Greeley's gardens on the south side of the Cache le Poudre; after that, for twelve years, none could be successfully grown for market. It made no difference whether they were planted in already cultivated gardens or on virgin soil, on irrigated or non-irrigated land, with or without applications of manure. On the north side of the river, in an area later known as the potato district, plants did better.[26]

A mid-sized, red-skinned potato, the 'Greeley Spud' likely descended from the 'Garnet Chile,' developed in 1857 by the Reverend Chauncey Goodrich of Utica, New York. Having studied the potato famine in Ireland, Goodrich hypothesized that after centuries of asexual propagation (by planting pieces of actual potatoes, not the potato seeds), the potato plant had become so weakened that it was no longer able to resist disease. Through the good offices of the US consul in Panama, Goodrich obtained specimens of a potato variety with rough, purple skin, which he named 'Rough Purple Chile,' reflecting its presumed country of origin. From the true seeds, whose seed balls look like globular groups of tiny green tomatoes, Goodrich grew his potato seedlings. He harvested the tubers (enlarged underground stems) and selected the best specimens, which he introduced as the 'Garnet Chile' cultivars. Four years later, in 1881, Vermonter Albert Bresse created 'Early Rose' from a self-fertilized seed ball of 'Garnet Chile.' Described by horticultural authority William Stuart as the first potato cultivar bred for commercial purposes in the United States, 'Early Rose' became the genetic stock for most subsequent potato cultivars and is sold today as an heirloom by specialized nurseries.[27]

The early Greeley potato growers soon discovered that no matter how carefully they cultivated and even if they grew potatoes on fields after an interval of five or more years, they had to cope with the Colorado potato beetle (*Leptinotarsa decemlineata* [Say])—an attractive, bright yellow-orange insect less than half an inch in length, with five brown stripes on each of

FIGURE 1.4. Spraying fruit trees, Grand Valley, 1911. *Courtesy*, Denver Public Library, Western History Collection, MCC 1475.

its two wings. Endemic to the Southwest where it feeds on native species belonging to the nightshade family (Solanaceae), including the potato, the beetle naturally attacks relatives in that family. The beetle lays eggs on the underside of leaves; both larvae and adults feed on the leaves, thereby injuring the plant and reducing or destroying its ability to grow tubers. During the 1850s, the beetle gradually spread eastward. To control the beetle, farmers initially had little choice except to hand-pick insects off their plants. By the time Greeley farmers began growing potatoes as a commodity crop, naturally occurring insecticides had become commercially available. The most effective was an arsenical poison known as Paris green, so-called for its use in eliminating rodents from Paris sewers and for its emerald green color. At first, growers used small whisk brooms or twisted bundles of straw to apply the poison to plant leaves; in time, they used hand pumps with spray nozzles and then horse-drawn mechanical spray machines.[28]

If the Colorado potato beetle caused the most severe injury to potato plants, a number of other diseases caused less damage but still affected yields or at least marketability. Captain Boyd described a fungus that appeared as rust, which "made the leaves thick and stiff, and undoubtedly destroyed the sap and prevented the leaves from carrying on their function." Most puzzling to growers was the occurrence of potato scab, which varied from field to field—generally affecting only the skin of the potato, not its yield, but with unappetizing blemishes. Today, we recognize that the common scab is caused by a microorganism, *Streptomyces scabies*.[29]

Searching for ways to prevent endemic nuisances from becoming epidemic outbreaks among their single-crop species, members of the Greeley Farmers' Club delved into the general topic of soil fertility. Reflecting on the absolute necessity of annually returning to the soil what harvested plants had removed, Captain Boyd invoked a familiar analogy: "If we keep drawing constantly from our bank account, without making fresh deposits, its depletion must come sooner or later, however large it was on the start." As such, Boyd reflected the ancient notion that developments in nature occur in cycles, a notion that has received increasing attention among present-day agriculturists and policymakers confronting dwindling natural resources and growing populations.

In Captain Boyd's understanding, all soils either contain the essential elements required to cultivate crops or receive elements through water and air. For the farmer, he explained, the question of soils can be reduced to: "Is there enough phosphoric acid, potash and nitrogen in the soil to supply the yearly drain of these substances?" All three of those elements exist in the soil but not always in the amount or form a given crop can readily absorb. On the labels of today's general-purpose fertilizers, those three elements are listed in the on-bag analysis as nitrogen (N), phosphorous (P), and potassium (K, from *kalium*, Latin for potash).[30]

Although Captain Boyd was no chemist, he likely had some familiarity with the work of the German agricultural chemist Justus Freiherr von Leibig, a near contemporary. Von Leibig had put forth the argument that the three inorganic elements (N, P, K) alone—inorganic because they are derived from inanimate minerals—enable plant growth. By the mid-nineteenth century, phosphorous and potassium derived from rocks had been mined and crushed into fertilizers, which were yet to appear commercially in Colorado. With regard to nitrogen, von Leibig hypothesized that if somehow one could isolate, collect, and distribute that element, then humus—organic material because it is derived from decomposed plants and animal manures—which does contain the three basic elements, would no longer be needed.[31] Boyd, of course, did not go that far; the age of synthetic fertilizers had yet to arrive. Instead, to make up for the drain on soil fertility and insufficient supplies of animal manures to revitalize the soil, Boyd advocated for increasing organic material through the cultivation of legumes, most notably alfalfa.

The name *alfalfa* comes from the Spanish transliteration of the Arabic *al-fasfasa*, meaning green fodder. The Roman Columella had called alfalfa medic because it was thought to have been introduced from Asia through Media in what is today northwestern Iran. He described medic as the best legume for improving the soil: its values include perennial, rapid growth that produces several cuttings within a single season, high quality as feed for livestock, and medicinal value to livestock (*On Agriculture* 2: 10, lines 25–26). Introduced to the South Platte Valley as a forage crop in the early 1860s, alfalfa met with a tepid reception; farmers found it extremely difficult to exterminate once planted, and most knew nothing about the long history of its nutritive and soil restorative qualities. Alfalfa gained popularity in Colorado only after popularized through newspaper articles by members of the Greeley Farmers' Club. Extensive cultivation of alfalfa around Greeley caused a veritable revolution in farming methods, according to Captain Boyd: "Not only does it [alfalfa] afford the means of making large quantities of manure, but it is found that, when turned under, it is even a better fertilizer than red clover [*Trifolium pratense* L., a short-lived non-native perennial]."[32]

In addition to adding organic matter to the soil, Captain Boyd noted that alfalfa plants absorb large amounts of nitrogen from the air, "if not directly through their leaves, which is doubtful, then through their roots." The nitrogen-absorbing function of leguminous root nodules, perhaps most evident on alfalfa roots, was yet to be fully understood. With exceptionally deep roots known to reach down as far as 15 feet, alfalfa brings up to the cultivated soil both phosphoric acid and potash, the latter required by potatoes more than by any other crop. Greeley farmers would learn from experience that the successful cultivation of potatoes required initial deep plowing and the maintenance of relatively loosely packed upper soil.[33]

Over the course of thirty years, the Greeley area became the largest potato-producing district in Colorado and the second largest in the nation.

Centered in Eaton, 8 miles north of Greeley, the district grew to about 30,000 irrigated acres. Although research on potatoes did not begin at the Colorado Agricultural Experiment Station until the late 1880s, growers had learned from experience that late-maturing cultivars did best; tubers develop only as soil temperatures moderate and daylight periods shorten. But neither stream flows nor reservoirs were sufficient for late-season irrigation. Some growers sought to supplement surface waters by digging wells and installing steam-powered pumps to bring up groundwater, but that, too, was not enough. As a result, the demand for supplemental water became the most pressing issue for Greeley-area farmers.[34]

To be sure, the question of water supply concerned not only the potato growers. Indeed, one cannot overemphasize that for farmers and non-farmers alike, no issue in Colorado and the West is more fundamental or contentious than water; there is never enough to go around. For the general reader, then, a brief explication on the rudiments of Colorado water law and its enforcement follows. Emerging during the 1870s, the Colorado system differed significantly from that based on riparian rights, which was common in Europe and the more humid regions of the United States where the right to use water belonged to the owners of stream banks. Under the Colorado system, the state owned the water. A farmer's right to access water went with the land to be irrigated, but that right was separate and distinct from title to the land. The right to use water, furthermore, was based on "first in time, first in right," the doctrine of prior appropriation.[35]

Colorado was the first state to include the doctrine of prior appropriation in its constitution in 1876. Six years later, in *Coffin v. Left-Hand Ditch Company*, a case appealed from Boulder County, the state supreme court confirmed that the doctrine of riparian rights was inapplicable to Colorado's semiarid conditions and that "the first appropriator of water from a natural stream for a beneficial purpose has . . . a prior right thereto, to the extent of such appropriation."[36] The state constitution (Article 16, Section 6) defined "beneficial use" in only the most general terms: "When the waters of any natural stream are not sufficient for the service of all those desiring the use of the same, those using the water for domestic purposes shall have the preference over those claiming for any other purpose, and those using the water for agricultural purposes shall have preference over those using the same for manufacturing purposes." To this day, precisely what those uses include remains a

FIGURE 1.5. North Poudre Canal, ca. 1890. *Courtesy*, Agricultural and Natural Resources Archive, Colorado State University, Fort Collins.

subject of debate. It is generally agreed, however, that the concept of beneficial use infers two practical consequences: first, water must not be wasted; second, if appropriated water is not used by its designated appropriator in a timely manner, that appropriator loses the right to use it.

As with the priority of usage doctrine, the process for enforcing water rights stemmed from experience with actual conditions, beginning with conflict over use of water from the Cache la Poudre River. In the hot, dry summer of 1874, Fort Collins farmers took out their appropriated share, which left Greeley farmers downstream with less than their share of water. Greeley farmers argued that they possessed prior rights to Fort Collins farmers, but there was no institutionalized process to resolve the dispute; in fact, this particular dispute was never resolved. Following adoption of the state constitution and proclamation of statehood, Greeley farmers under the leadership of Captain Boyd started the debate that, over a period of several years, would engage the entire spectrum of farmers and civic leaders from communities dependent on the waters of the South Platte and its tributaries; the debate culminated with deliberations of the select legislative committee chaired by Captain Boyd. That committee drafted the enforcement legislation approved by the legislature in 1879 and 1881. One would like to believe that Captain

Boyd's success in forging agreement on such a controversial issue derived at least in part from his having adopted the cooperative spirit pioneered by Nathan Meeker.

For purposes of appropriating and administering Colorado waters, state legislation provided that the state be divided into drainage divisions that corresponded to the hydrographic basins of the principal rivers; the divisions were further divided into districts. At first, only nine districts were defined, all in the South Platte Valley, to make up division one. In time, the state would be divided into seven divisions, eventually subdivided into eighty districts. For each district, the governor was authorized to appoint a water commissioner to assign priorities of appropriation as provided in written records, with the process for obtaining such records also established. To administer the process, the legislature established the Office of the State Engineer; its major responsibility was to accurately measure stream flow—a particularly sensitive task given the lack of historical data combined with the gross overestimation of stream flow by those who held the water rights.[37]

The legal and administrative aspects of water management hardly kept up with the rapid development of canals and ditches, which by the mid-1880s extended far beyond the network using the Cache la Poudre River. Farmers greatly expanded the amount of irrigable land throughout the South Platte, Arkansas, and Rio Grande Valleys—causing changes in the makeup of soils as well as in the overall agricultural economy. During the Hayden survey of 1869, scientists had observed that downstream flows, where irrigation waters had been diverted upstream, were greater than on their first trip two years earlier. Not fully understood until the late 1880s, return water—now called return flow—resulted from irrigation water that percolated its way back to the stream. Farmers claimed new water rights and built new canals so they could use that water. Captain Boyd and others had already recognized that excessive irrigation could remove valuable nutrients from the soil.[38]

The Union Colony as a corporate body had built and maintained the Greeley-area irrigation system but soon turned ownership over to participating irrigators. As demand for irrigation water increased, private investors saw opportunities for profit in constructing and operating ditches and, in addition to speculating on irrigable land, in charging farmers royalties for use of the waters. Among the largest such private equity firms was the London-based Colorado Mortgage and Investment Company, developer of the Larimer and

Weld Canal designed to irrigate 60,000 acres—one-third of the land owned by the company—and the High Line Canal, which took water out of South Platte Canyon for a projected 124 miles, meant to cover 200,000 acres. The prospect of having non-farmer, non-local investors controlling Colorado water generated widespread opposition from farmers. Irrigators appealed to the state government for help; their most notable success was the Colorado Supreme Court decision in *Wheeler v. Northern Colorado Irrigation Company* (1887). The court determined that ditch companies were common carriers, with no property interest in the water; they could levy reasonable charges for transporting water but could not assess royalty fees on the water itself.[39]

Farmer opposition to the ditch companies had contributed to a more general criticism of land speculators, monopolists, and wholesalers. Such opposition was given context and direction by a succession of mutual benefit agricultural associations starting with Colorado chapters of the National Grange of the Patrons of Husbandry followed by the National Farmers' Alliance and Industrial Union and after the turn of the twentieth century by the National Farmers' Union.

The Grange began as a secret fraternal organization in upstate New York in 1867 for the expressed purpose of improving farm life and combating the perceived evils of modernism. The period following the Civil War was one of rapid economic expansion, for agriculture made possible in part by technological advances such as the mechanical reaper and the steel ploughshare. The federal government provided financial incentives for big corporations—most notably, the railroads pushing westward—and for individual farmers through preemption acts, although loopholes in those acts led to land speculation that countered legislative intent. Grangers started the first Colorado branch in Boulder in 1874 and quickly established sixty-nine other branches, reaching a peak of eighty-five branches with a publicized membership of 2,390 in 1888. Because the Grange leadership insisted on remaining non-political, the more militant National Farmers' Alliance became more attractive to disaffected farmers.[40]

Like the Grange, the National Farmers' Alliance began as a secret fraternal organization; the tradition of secrecy went back centuries to lay Christian associations and is still prevalent among some religious confraternities. Founded by farmers and ranchers in central Texas during the 1870s, the alliance spread first through the South and Midwest. Its initial goal: to establish member-owned cooperatives as a way to eliminate the need for intermediaries such as wholesalers and distributors in an effort to give farmers control of prices for produce and livestock as well as for the farm supplies and equipment they bought. By the time of its annual convention at Cleburne near Dallas in 1886, the alliance had expanded and politicized its mission, petitioning the federal government to regulate railroad rates, impose punishing taxes on land speculators, and compel banks to increase the availability of farm credit. More specifically, at its convention in St. Louis two years later, leaders circulated a proposal for the federal government to lend grain producers up to 80 percent of the value of their crops at a low interest rate, pay for grain storage, and allow producers to reclaim their stored grain for their own use or for sale when prices were higher. The intent of this proposal—to stabilize and protect farm income and prices—would eventually underlie the Commodity Credit Corporation Act, part of the New Deal's legislation for relief, recovery, and reform.[41]

In 1888, National Farmers' Alliance organizers began recruiting members in Colorado, starting with irrigators in the Arkansas River Valley; next, in eastern Colorado, where crop failures had led to widespread anger by farmers who had been misled about the climate by unscrupulous agents representing the railroads; and then in northern Colorado to organize irrigators dissatisfied with the Grange, in particular, its ineffective opposition to the non-agricultural owners of the irrigation systems. By 1891, the alliance claimed 15,000 members throughout the state.[42] But that was not all. During the summer of 1890, the National Farmers' Alliance had made the strategic decision to join forces with the Knights of Labor, which represented miners and industrial workers, another disaffected element of the population.

Officially named the Noble and Holy Order of the Knights of Labor—cynically or otherwise, an acknowledgment that its roots were not simply economic—the Knights of Labor had also begun as a secret society but more like a guild, started in 1869 by skilled garment workers in Philadelphia. By the time the Knights of Labor reached Colorado, the organization had become very successful in organizing lesser skilled laborers such as miners and factory workers. By the standards of the day, the Knights of Labor espoused radical ideas—the eight-hour day, abolition of child labor, women's suffrage, and a graduated income tax with those having the lowest income paying the lowest percentage of tax.[43] Because such ideas generally did not sit well

FIGURE 1.6. Plow pulled by mules, undated. *Courtesy*, Agricultural and Natural Resources Archives, Colorado State University, Fort Collins.

with independent, profit-minded farmers, their alliance with labor did not last—but not before the National Farmers' Alliance and Knights of Labor banded together to found a new political party called the Independent Party that in 1892 melded with the People's Party, a nationwide alliance of disparate groups and individuals united in opposition to the major political parties for their failure to control unfettered capitalism.[44]

Although a full discussion is beyond the scope of a study focused on the cultivation of the land, the People's Party, better known as the Populist Party, fared better in Colorado than in any other state, and part of its success has been attributed to support by the National Farmers' Alliance and its members. More to the point: the price of silver, once the lifeblood of the state's economy, had been dropping precipitously as the nation and others moved toward the gold standard; neither Democrats nor Republicans were especially interested in reviving silver. As a result, the People's Party adopted the cause of "free and unlimited coinage of silver." The party's standard-bearer was

Davis H. Waite, an eccentric upstate New Yorker, trained in the law, who had moved west in search of opportunities. Eventually, he reached Leadville to prospect; after a year he moved to Aspen where he returned to the law, representing mine workers against mine owners. In 1891, Waite started the *Aspen Union Era* to vent his opposition to the wealthy and the so-called monopolists in favor of the laborers and the common folk, which brought him statewide notoriety and helped launch his successful bid to become the nation's first and only Populist Party governor (1893–95).[45] Although oriented toward labor, the Populist Party platform did call for member-owned agricultural cooperatives, state ownership of irrigation systems, and nationalization of railroads. Historian James E. Wright has documented that Waite carried only three of the ten counties with the most National Farmers' Alliance members. His support from industrial workers, his promotion of their economic and social agenda, and his ineptitude in governing—a characteristic of ideologues in perennial opposition—alienated both business and agricultural interests.[46]

The agriculturists most fearful of populism were the more successful producers of commodity crops. No commodity crop depended more on unskilled laborers than the sugar beet. Similarly, none depended more on irrigation than the sugar beet. Even prior to widespread irrigation, Colorado farmers showed interest in the high-sucrose content of the long, fleshy, and whitish to reddish-purple root of the sugar beet (*Beta vulgaris* L.), a member of the same family as the garden beet and chard. Fifty-Niner turned farmer Peter Magnes brought seeds from Illinois and planted the first sugar beets in the South Platte Valley in the early 1860s. At about the same time, L. K. Perrin started to grow sugar beets on his farm along Clear Creek near Golden. Both farmers sent samples to Jacob F.L. Schirmer, metallurgist and chemist at the Denver Mint; his analyses showed that the sugar content of Colorado-grown beets was significantly higher than that of beets grown in his native Germany. That prompted William Byers, ever the publicist, to editorialize that "the soil of Colorado has no superior in the world for producing the sugar beet." Lamenting that no processing facility existed anywhere in the United States, Byers wrote that such a facility would be a profitable investment. He cited as evidence the tremendous growth in sugar beet production and beet sugar consumption in France where, unbeknown to Byers, an entrepreneurial sixteenth-century French farmer was recognized as the first person to extract sugary syrup from the beet. His experimenting far from Paris attracted the attention of the monarchy, always interested in replacing goods from overseas with those produced domestically.[47]

In the Union Colony's first annual report (1871), William Pabor exuded optimism about the future of sugar beets in Colorado. Unlike locations in the Midwest, Greeley's soil and climate were considered ideally suited for sugar beets. Thousands of acres could be profitably grown, provided that some farsighted entrepreneur was willing to invest a sizable fortune in a processing facility. It took nearly thirty years before such facilities were built, enabling sugar beets to be grown as a sugar-producing commodity crop. Some small diversified farmers, meanwhile, experimented with sugar beets as forage for dairy herds. Once processing facilities began operating, refiners sold beet pulp and beet tops as supplements in finishing rations for livestock, contributing to the development of an extensive and entirely new business of sheep and cattle feeder lots, which still grace Greeley and environs with a noticeable odor.[48]

Beet fever struck Colorado during the early 1890s, when farmers—especially in the Arkansas River Valley—were experiencing the harmful effects on the soil of successive plantings of grain crops in the same fields. They discovered that sugar beets grown in rotation with grains helped break disease cycles and balance crop nutritional needs, producing higher yields. In addition, cultivation of beets was relatively easy and inexpensive, requiring minimal irrigation and offering higher profits than grains. What ultimately attracted financiers to invest in beet processing facilities was the economic development drumbeat of the Denver Chamber of Commerce combined with encouragement from the US Department of Agriculture (USDA). Test results from Colorado Agricultural College validated the claims of high sucrose content of Colorado-grown sugar beets.[49]

Aside from funding and technical assistance provided by the USDA, the college's combined faculty and staff would exercise the single most important influence on the practices of agriculture in Colorado. By 1870, farmers in Larimer County had successfully lobbied the territorial legislature to enable the creation of an agricultural college in Fort Collins, their county seat; no funds were immediately appropriated. Three years after statehood, the legislature and governor accepted the provisions of the Morrill Act. Signed by President Abraham Lincoln in 1862, the act authorized the federal government to grant to each state unappropriated land that the state could sell,

FIGURE 1.7. Young farmer with load of sugar beets, 1910–27. *Courtesy*, Denver Public Library, Western History Collection, MCC3210.

invest the proceeds in a perpetual fund, and use the earnings of that fund to support a college for agriculture and the mechanic arts—leaving it to the discretion of each state how best to "promote the liberal and practical education of the industrial classes in the several pursuits and professions in life."[50] That flexibility would lead to the interminable debate between those who held the narrow view of the agricultural college as teaching students how to cultivate the soil versus those who believed the college should help students understand the basic principles of science so they could intelligently apply those principles to farm or ranch. The latter view meant no disdain toward agricultural practitioners. Instead, it reflected the belief that in a liberal democracy, everyone needs to know how to read, write, and count and to know something about the past and about plants and animals. Members of the Colorado Board of Agriculture, which served as the governing body of the agricultural college, tended to favor the narrow view. Board members

David Boyd and Bryant S. LaGrange favored a vocational-technical college for farm youths. Elijah E. Edwards, the college's first president (1879–82) and an ordained Methodist minister, held the broader view. He resigned after less than three full years in office.[51]

Prior to appointing the first president, the Board of Agriculture had employed Ainsworth E. Blount to manage the college farm, which had existed since 1870, cultivated by volunteers from the Grange. Like many agricultural scientists of his time, Blount was the son of missionaries—in his case, ministering among the Cherokees. He had earned a bachelor's degree in the liberal arts at Dartmouth College, served in the Union Army, and presided over a small women's college in Tennessee before arriving at Colorado Agricultural College. President Edwards promoted him to professor of practical agriculture, with teaching and research responsibilities. As head of the new Experimental Department, Blount tested dozens of varieties of grain, grass, and vegetable seeds for their adaptability to Colorado's soil and climate. Especially interested in adapting corn from the eastern states to Colorado, he selected ears that ripened early and contained the most kernels, distributing samples to farmers to obtain further information on yields under a variety of conditions. Although unexciting, his experimental work did reveal an appreciation for the need to maintain soil fertility. To fully demonstrate the advantages of careful rotation, he used no fertilizers, never allowed a given crop to succeed itself, always planted grains following market garden crops, and recognized the benefits of nitrogen-fixing alfalfa to the soil.[52]

Among the early agricultural faculty, Elwood Mead surpassed them all for his contributions to agricultural science and its applications. A native of rural Indiana, he was attracted to Colorado by his undergraduate mentor at Purdue University, Charles L. Ingersoll, who became president of Colorado Agricultural College in 1882. Mead started as a professor of mathematics and physics, then took a leave to earn a bachelor's degree in civil engineering at Iowa Agricultural College (now Iowa State University) and a master's degree at Purdue. He returned to Colorado to a dual appointment as assistant to the state engineer and professor of irrigation engineering, the first such academic position in the United States. Although assigned little space on the college farm by his colleague Blount—academic jealousies are not new—Mead did conduct experiments regarding the duty of water, that is, the volume of irrigation water required for any given commodity or market crop. Mead opposed outside investor ownership of irrigation systems, which made him popular among northern Colorado irrigators and enabled him to critically assess the Colorado system of water management. Similar to Nathan Meeker, Mead believed that irrigation fulfilled a social as well as an economic purpose, which was to ensure that the arid West would become and remain the home of small-scale farmers—a position that gained popularity after the turn of the twentieth century. In 1890 he left Colorado to become Wyoming's first state engineer, and he ended his lengthy career as director of the US Bureau of Reclamation (1924–36).[53]

Although overshadowed by botanists Charles Bessey at Nebraska and Aven Nelson at Wyoming, James Cassidy did introduce landscape gardening to Colorado. An Englishman by birth, Cassidy had worked as a florist at the Royal Botanic Gardens before immigrating to the United States, earning a degree at Michigan Agricultural College (now Michigan State University) and working there as florist and gardener before moving to Colorado in 1883. On the college farm, he ran test plots on garden vegetable, fruit, and grain varieties and conducted the first tests on potatoes.[54] To improve yields, he tried unsuccessfully to cross the small native potato (*Solanum jamesii* Torr.) found in southwest Colorado with existing cultivars. He did confirm the experience of early market gardeners that in Colorado, "feeding the soil" is "necessary, desirable, and profitable." Cassidy experimented with four Colorado-manufactured fertilizers—bone superphosphate, plaster, Missouri clay kalsomine, and bone meal. Although relatively expensive, bone meal, a mixture of crushed and ground animal bones, showed the best results. Synthetic fertilizers were yet to be developed.[55]

As Colorado became better connected to the rest of the nation, researchers at the experimental farm were faced with protecting agriculture from non-native insect pests and noxious plants, the most notorious being Russian thistle (*Salsola kali* L.)—often called tumbleweed because it is blown in masses by late summer winds. Not a true thistle, Russian thistle had been introduced inadvertently with flaxseed from Eastern Europe to South Dakota in the mid-1870s; from there, it spread rapidly across the plains states. Less apparent but also damaging were insect pests including the cabbage butterfly (*Pieris rapae* L.), which preyed on vegetables such as cauliflower and Brussels sprouts, and the coddling moth (*Cydia pomonella* L.), which fed on fruit, especially apples and pears. With the cooperation of Colorado farmers

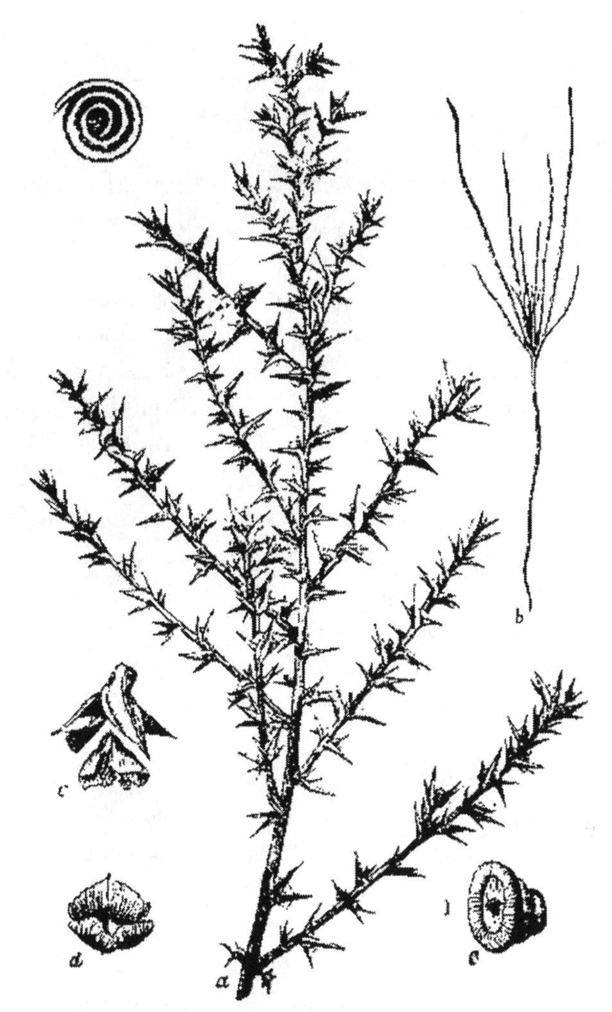

FIGURE 1.8. Russian thistle, 1894. *Courtesy*, Agricultural and Natural Resources Archive, Colorado State University, Fort Collins.

who mailed him specimens, James Cassidy began systematically to identify the state's insect pests as well as ways to eliminate them. He recommended two types of insecticides: arsenical poisons that kill through ingestion, for insects that eat plant structures; and alkali, acid, or oil mixtures that kill through contact, for insects that feed on viscid leaf surfaces. Although he wrote nothing about beneficial insects that prey on destructive insects, he did observe that plants in "unthrifty" condition as a result of soil exhaustion were most vulnerable to pest attacks—more recognition that healthful soil produces healthful plants.[56]

Cassidy's decidedly more rigorous research approach reflected the institutional transition from college farm to experiment station that followed passage of the Hatch Act, signed by President Grover Cleveland in 1887. The act authorized and appropriated monies for the creation of a system of state agricultural experiment stations, with each station operated separately according to the conditions and needs of that state. Each station was to conduct original research or verify experiments on the physiology of plants and animals; the diseases to which they are subject, with remedies for same; the chemical composition of useful plants; the advantages of crop rotation; the capacity of new plants or trees for acclimation; analysis of soils and waters; the chemical composition of manures; the adaptation and values of grasses and forage plants; the composition and digestibility of foods for domestic animals; scientific and economic questions involved in the production of butter and cheese; and other research or experimentation bearing directly on agriculture in the United States.[57]

For their inaugural experiment station project, Cassidy and chemist David O'Brine undertook research on the sucrose content of sugar beets. They had chosen a quarter-acre plot consisting of clay-loam soil held in clover for the prior three years and plowed the previous fall. From the USDA they obtained seeds of four beet cultivars, using a mechanical seed drill to sow them in rows 3 feet apart. Although yields of beets by ton and sugar in pounds varied by cultivar, all compared favorably with results from other states. Three years later, in 1890, two Colorado Agricultural College graduates, Frank L. Watrous at the Arkansas Valley substation in Rocky Ford (established 1888) and H. H. Griffin at the new San Luis Valley substation in Del Norte, planted test plots. Their harvests together with specimens from farmers in both areas went to Fort Collins for sucrose testing. In addition, Watrous reported on wheat growers considering sugar beets as rotation crops to mitigate soil deterioration caused by continuous cropping.[58]

Independent of efforts east of the Rockies, agriculturists on the Western Slope had taken interest in cultivating sugar beets between rows of peach trees. In 1890 or 1891, Grand Junction orchardist Henry R. Rhone persuaded his friend, US senator Henry M. Teller, to exercise the traditional congressional perk of requesting seeds for constituents from the secretary of agriculture. Rhone then sent harvested samples back to Washington for testing. Favorable reports encouraged his fellow orchardists to interplant beets. Mesa County growers entered into an agreement with the Utah Sugar Company at Lehi, about 250 miles west of Grand Junction. The company supplied seed and sent a representative to oversee planting on about 75 acres; growers shipped their beets to the Lehi refinery by rail. Higher prices received for orchard fruits and difficulties encountered in interplanting dampened enthusiasm for growing beets as a major cash crop.[59]

The coming of the Spanish-American War and the anticipated interruption in the sugar supply revitalized investor interest in producing sugar domestically. That interest was spurred by passage of the Dingley Tariff Act of 1897, continuing the tradition in place since the founding of the republic to protect domestic sugar producers. Western Slope promoters obtained pledges from farmers to plant 3,500 acres in sugar beets for at least three years and attracted financial incentives from the Mesa County Commission. In 1898, a group of investors led by mining magnates Charles Boettcher and John F. Campion built Colorado's first sugar beet processing plant in Grand Junction. Although that factory closed after one year because of issues with growers, Boettcher and Campion went on to found the Great Western Sugar Company, starting with refineries in Loveland (1901), Eaton and Greeley (1902), and Windsor (1903). In the South Platte Valley, Great Western Sugar Company helped underwrite expansion of irrigation works, even constructing and operating its own railroad, Great Western Railway, to better connect producers with the refineries. Thanks to stiff protective tariffs and farmer enthusiasm, investors would build twenty refineries in Colorado between 1900 and 1920.[60]

With tremendous growth in sugar beet acreage (about 80,000 acres in the South Platte Valley by 1910), both refiners and growers required large numbers of workers—in particular unskilled workers—to do the strenuous manual work of cultivating the fields. Since most local residents either had jobs or declined to take those back-breaking, low-paying jobs, growers recruited in the Midwest and abroad. Their greatest successes were in recruiting and hiring Germans from Russia, also known as Volga Germans, mostly poor farmers who had fled czarist Russia beginning in the 1880s to escape famine and political suppression of ethnic minorities. It was only after the Mexican

FIGURE 1.9. Great Western sugar beet factory, Fort Collins, 1905. *Courtesy*, Agricultural and Natural Resources Archive, Colorado State University, Fort Collins.

Revolution of 1910, when Mexicans were fleeing economic instability, that Mexican laborers began replacing Germans from Russia, many of whom had become growers themselves. The Museo de los Tres Colonias in Fort Collins preserves the story of Hispanic life and working conditions in the sugar beet industry during that time.[61]

Although he had dabbled in sugar beets, Henry Rhone was first and foremost an orchardist. He had been attracted to the rich soil and plentiful water of the Grand Valley, part of the last remaining tribal lands in the United States opened to settlement in 1881. It is a good bet that Rhone and his fellow agriculturists took advantage of preemption legislation to purchase land and got together to build an irrigation system for diverting Colorado River waters to their respective farms and orchards. A variety of fruit trees—apples, cherries, plums, peaches, and pears—were introduced, with peaches planted the most. Recognizing the commercial potential for orchard fruits, D. S. Grimes, a Denver nurseryman and president of the recently formed Colorado State

FIGURE 1.10. Apple orchard with regulation-size wooden boxes. *Courtesy*, Agricultural and Natural Resources Archive, Colorado State University Libraries, Fort Collins.

Horticultural Society, established a nursery near Grand Junction in 1883. During the next year William Pabor, in his capacity as agent for a Denver bank, established the Fruita Town and Land Company: 520 acres plus accompanying water rights, an attractive package for prospective settlers.[62]

In the course of his travels through Colorado, Pabor had found a variety of small fruits—red, black, and yellow currants, raspberries, strawberries, whortleberries, and June berries; also "wild plums of larger size and as delicious flavor as one-half of our cultivated varieties." If native fruits could be "brought to the light of cultivation," he mused, they would become "jewels in the horticulture of Colorado."[63] In fall 1881 two farmers, Enos T. Hotchkiss and Samuel Wade, scouting for homesteads on soon-to-be-opened Ute tribal land, found buffaloberry (*Shepherdia canadensis* [L.] Nutt.) and thornapple (probably *Datura wrightii* Regel) in the North Fork Valley of the Gunnison River. Their discoveries encouraged them to believe that cultivated small fruits and orchard fruits could be successfully grown; in spring 1882 they returned with saplings—apples, pears, apricots, and peaches—as well as blackberry and grape root stock for their homesteads located on the sites of present-day Paonia and Hotchkiss. By 1900, fruit growing had spread to other parts of Delta County as well as to Montrose County, complementing more than competing with orchards in Mesa County. Beneficial to the fruit market, when early frost damaged North Fork orchards, which naturally occurred, it generally spared the Grand Valley and vice versa.[64]

Long before the opening of the Western Slope to settlement, Coloradoans had recognized their dependence on forests for grazing, timbering, and, of course, water. In 1876, the state constitution (Article 18, Section 6) called for the state legislature to enact laws to "prevent the destruction of, and to keep in good preservation, the forests upon the lands of the state, or upon lands of the public domain." Farmers and orchardists dependent on mountain runoff generally welcomed President Benjamin Harrison's executive order of October 16, 1891, setting aside 1.2 million acres of unappropriated federal land in northwest Colorado. The White River Plateau Timberland Reserve was the first such reserve in Colorado and the second in the nation. In contrast, agriculturists at higher elevations dependent on the forest for timbering, grazing, hunting, and fishing vehemently opposed the presidential order, setting up the first major public debate over conservation in Colorado. Even more significant, by passage of the Forest Reserve Act of March 3, 1891, the US Congress and the president had set into motion a fundamental change in federal policy: from selling public lands to encourage settlement and earn income for the federal government to keeping public lands in the public domain for the common good and to still earn income for the government. That change in federal land policy would fundamentally affect agriculture in Colorado and the West.[65]

But not before the federal government, seeking to promote settlement in the West, had granted the railroads broad swaths of land beyond the reaches of irrigation. Along their tracks, the railroads laid out town sites and hired agents to sell real estate; the greater the population, the greater the income for the railroads. During the wet years of 1868 and 1869, when the Kansas Pacific line was nearing completion through the grasslands of eastern Colorado, railroad agents promised immigrants—mostly from eastern Kansas and eastern Nebraska—that they could successfully farm using their accustomed methods of cultivation. Agents peddled the false notion that precipitation would increase as the grasslands gave way to cultivated fields ("rain follows the plow"). As normal cycles of drought, grasshopper plagues, and severe winters occurred, most of those immigrants abandoned the land. For the few who remained as settlers, survival depended on cultivation that preserved soil moisture without supplying additional irrigation, the method that became known as dry-land farming.[66]

Whether on the eastern plains, in the central mountains, or on the western plateau, the general pattern of cultivated lands in Colorado appeared pretty much set by the turn of the twentieth century. Between 1870 and 1900, the number of farms in Colorado had increased from 1,738 to 24,700; acres under cultivation had grown from 95,594 to 2,273,968, of which irrigated acres increased from a few hundred to 1,611,271—making Colorado first in the nation for the number of acres under irrigation and second in both the number of irrigators and the value of irrigated crops. The 1900 US Census identified 118 commercial farming districts nationwide that were leading in vegetable production, of which three were in Colorado: the Denver district (Arapahoe and Jefferson Counties) producing for the local market in season, mostly notably cabbages, cucumbers, tomatoes, sweet corn, and sweet peas; the Rocky Ford district (Fremont, Otero, and Prowers Counties) with ten-year growth unsurpassed by any district, primarily in muskmelon but also in tomatoes and watermelon; and the Greeley district (Boulder, Larimer, and Weld Counties), especially Weld County, producing more than half of the state's potatoes for shipment to distant as well as local markets. For Weld County the 1900 census reported 251,307 acres under cultivation on 2,002 farms compared to the 1870 census that recorded 9,715 acres on 105 farms.[67]

Whether in the south-central part of the state, the Arkansas and South Platte Valleys, the eastern plains, the central mountains, or the western plateau, the general pattern of agricultural land in Colorado appeared established by the turn of the twentieth century. Statewide, however, unfettered growth in farming and grazing posed considerable challenges—some recognized and others yet to be understood—for both public and private stewards of the land and water. Mexican sheep in south-central Colorado, meanwhile, had given way to merino and other breeds that numbered more than 2 million animals in 1899. As with farming, ranching had attracted ideologues as well. William Pabor reported visiting a young couple at their ranch near Kiowa. Recent arrivals from the East, they were smitten by the "novelty and charm" of the shepherd's life. It was not uncommon, he sentimentalized, to find such families "temporarily occupying a dug-out [earth lodge], and having with them a piano and other evidences of culture and refinement."[68]

2

Making the Land Flourish

At the turn of the twentieth century, farming in Colorado prospered wherever irrigation waters were plentiful, just as William Pabor had envisioned. Along the Front Range, market gardens, grain fields, and livestock operations dotted the countryside. The situation on the eastern plains was far different. By the mid-1890s, the few intrepid farmers remaining had learned that to survive, they needed to use water sparingly and conserve moisture by digging wells, building small reservoirs, and cultivating in unaccustomed ways to grow grain crops.

It is too bad that we know so little about Joshua William Adams, an 1898 graduate of Kansas Agricultural College. He had worked summers on the Colorado Agricultural College demonstration farm at Cheyenne Wells, anticipating full-time employment after graduation; but when the farm was forced to close the year he graduated, he started his own farm 16 miles south of town. Thanks to a legislative appropriation ten years later, the college reopened the site, converted it into an experiment substation, and appointed Adams as superintendent. Although he received no salary, Adams lived there

https://doi.org/10.5876/9781646422050.c002

rent-free; he was also allowed to move his own cattle and other farm possessions to the substation and consume or sell for his own benefit any agricultural product grown at the substation after recording test plot yields and related measurements. His tenure continued into the mid-1920s. He built up a dairy herd, starting with a single purchased Holstein bull and two cows; dug the first pit silos for holding grain on the eastern plains; and built a dairy barn and other substation buildings using sun-dried bricks he had formed out of the clayey soils. Adobe as building material seemed eminently practical for the treeless country.[1]

Joshua Adams was a proponent of small-scale, primarily subsistence farming. He urged each of his neighbors to be satisfied with a small dairy herd, a few hogs and poultry, and a well-protected kitchen garden with access to well water. Cash income could be derived from the sale of surplus milk, transported relatively inexpensively in condensed form. Considering the frequency of crop losses due to inclement weather and the fact that grain prices were beyond grower control, Adams argued that making grain one's principal dry-land crop was far too risky. What made his commentaries prescient, if not extraordinary for his time, were his specific recommendations on techniques of soil conservation. "Great care should be taken to protect the land from blowing," he wrote. Wind erosion "not only ruins growing crops, but injures the soil." He recommended leaving stalks or stubble in the field after harvest, making ridges and furrows by using the plow in the fall, and cultivating in early spring to prevent weeds from growing. Adams was credited with introducing contour plowing—plowing along the elevation—to the eastern plains; such plowing helped prevent erosion by controlling runoff during cloud bursts, the sudden violent rainstorms typical of semiarid country.[2]

Joshua Adams may have been ahead of his neighbors with regard to adopting agricultural techniques that helped conserve soil moisture, but interest prevailed nationally in conservation, at the time understood to mean the fullest possible use of renewable resources in such a way as to ensure their permanence. In his renowned first annual message to the US Congress, President Theodore Roosevelt set out a conservation agenda that spoke directly to farmers and ranchers in the West. The agenda was divided into forest conservation and water conservation, reflecting the expertise of his two most trusted conservation advisers—US Department of Agriculture (USDA) forester Gifford Pinchot and US Geological Survey hydrographer

Frederick H. Newell. Protecting forests while exploiting them had long been of economic, military, and scientific concern; but vivid reports of denuded hillsides, wasted timber, and degraded watersheds around mining camps—especially in the Black Hills—had elevated the forests to highest priority on the president's conservation agenda and emboldened Pinchot to undertake the year-in and year-out scheming that led to federal administration of the forest reserves and to the establishment of the Forest Service within the USDA. Placement there rather than in the Department of the Interior reflected Pinchot's approach to forestry as the science of managing stands of trees as agricultural crops, the cultivation of which led him to become embroiled with those who used and misused the forest. Historian G. Michael McCarthy recounted an early Colorado story of opposition to the Forest Service. On December 4, 1905, ranchers crowded into the ballroom of the Glenwood (Springs) Hotel to confront Pinchot, by then chief forester, on the subject of grazing permits and the imminent imposition of grazing fees; they were followed by farmers who complained that Forest Service regulations were as odious to them as to the stock growers. Pinchot, ever the smart, self-confident eastern conservationist, listened attentively for an entire morning. For an hour and a half in the afternoon he charmed his listeners, presumably citing government reports on the White River Plateau Timberland Reserve, where overgrazing had nearly eliminated ground cover and destroyed young trees needed to regenerate and maintain the forest. His message was clear and direct: the purpose of federal management was to ensure that grazing could continue indefinitely; user fees to support such management would go into effect as scheduled. At the conclusion, with no sounds of disapproval, the ballroom remained silent. That evening Pinchot left by train for Washington, DC, with outwitted ranchers eventually relenting on the forest issue. Plenty of opportunities would arise and still do to stir up state and local opposition to federal management, but there is little opposition to federal underwriting of water projects that also began under President Roosevelt.[3]

In that first annual message, Roosevelt had argued that conservative forestry alone could not ensure an adequate year-round supply of water for the arid West; only the federal government could command the resources necessary to undertake the massive projects required to regulate stream flows and save spring runoffs for beneficial purposes. The Reclamation Act of 1902 authorized the secretary of the interior to construct storage reservoirs as well as canals and other means for controlled water distribution in the sixteen western states and territories and to pay for them through the sale of public lands and assessments on water users over a ten-year period. As would happen so often in the federal government's relationship with the states, congressional appropriations in time would cover nearly all reclamation costs; the states retained control of the waters. In 1907 the president appointed Frederick Newell as the first director of the Reclamation Service, later renamed the Bureau of Reclamation.[4]

One of the agency's first projects occurred on the Western Slope, thanks in great part to the surveys done since the late 1880s by A. Lincoln Fellows, a US Geological Survey hydrographer stationed in Colorado, and to advocacy by Representative John F. Shafroth (Silver Republican–CO). The Uncompahgre Reclamation Project included construction of a 12-mile-long canal—6 miles tunneled—to divert waters from the Gunnison River to the Uncompahgre River, supplying water to agriculturists in Montrose and Delta Counties. Reclamation engineers estimated that the completed project would provide supplemental waters to 130,000 acres already under cultivation and to another 40,000 acres of proposed irrigated land. By 1925 the project covered 61,637 acres, a fraction of the total irrigated acres statewide.[5]

President Roosevelt's support of the Reclamation Act reflected his deep personal commitment to the West, effectively an endorsement of Major John Wesley Powell's thesis that by supplying irrigable water to the arid West, the federal government would ensure that the region would develop and remain a place of small farms and small towns. But the president's call for the conservative use of forests and waters was part of a broader socioeconomic agenda set forth in the report of his Commission on Country Life (1908–9): "to develop and maintain on our farms a civilization in full harmony with the best American ideals." To do so meant, "first of all, that the business of agriculture must be made to yield a reasonable return to those who follow it intelligently; and life on the farm must be made permanently satisfying to intelligent, progressive people." Toward that end, the report included three principal recommendations: an exhaustive survey of rural life, the organization of a nationalized agricultural extension service, and a publicity campaign for rural progress. The report has been derided by some as the work of urbane easterners seeking to restore the charms of rural life

FIGURE 2.1. Irrigation, Delta County, 1928. *Courtesy*, Agricultural and Natural Resources Archive, Colorado State University Libraries, Fort Collins.

and by others as a cynical government effort to make farming more efficient, keeping food prices low for restless industrial workers. In retrospect, the commission's report served as the first step in a national effort to develop "a system of self-sustaining agriculture."[6]

Such a system depended, first and foremost, on the restoration and maintenance of soil fertility. According to the commission, historically, farming in America had been "largely exploitational, consisting of mining the virgin soil" and lasting no longer than two generations before causing "results pernicious to society." The vitality of rural communities depends on fertile soils; poor land equals poor people. The commission proceeded to praise those thoughtful farmers committed to "self-perpetuating agriculture," contrasted with more tradition-bound farmers abandoning depleted soils for new land or remaining to end their lives in "poverty and degradation." To lead and demonstrate the way to "scientific agriculture," the commission looked to the USDA and the land-grant colleges.[7]

Not to be overlooked, underlying the commission's recommendations was a certain spiritual conviction about the sacredness of the land, shared alike by Roosevelt-era scientists and non-scientists, churchgoers and non-churchgoers. Its leading advocate happened to be the commission's chairman, Liberty

Hyde Bailey, a keen naturalist, dean of agriculture at Cornell University, and prolific writer for both academics and the general public. In 1915 he published arguably his most profound and influential book, *The Holy Earth*; it has been reprinted several times, most recently in 2009. Bailey's thesis on the earth's holiness and humanity's obligation to serve as its steward has been resurrected in recent years by a combination of theologians and conservationists under the title of eco-spirituality, turned into an academic discipline known as eco-theology. Bailey's thesis would provide a philosophical basis for Governor Dick Lamm's plan for economic and community development within nature's limits—a moral justification for the burgeoning natural and organic food movement—but we are getting ahead of our story.

As a progressive, an evolutionist, and a person of faith, Liberty Hyde Bailey interpreted Judeo-Christian scripture from a decidedly agrarian point of view. He understood the story in Genesis 2:15 ("the Lord God took man, and put him into the Garden of Eden to dress it and to keep it") to mean that humans are given dominion over the earth but not the authority to devastate it. Before attending college and still on the family farm in western Michigan, Bailey had read Darwin's *Origin of the Species*. This helps explain his later interpretation that we are placed on earth "not as superior to nature but as a superior intelligence working in nature as a conscious and therefore as a responsible part in a plan of evolution, which is a continuous creation."[8] The farmer, then, performs a unique role: "to guard and to subdue the surface of the earth" as the "agent of the divinity that made it." Like Aldo Leopold, Wendell Berry, Wes Jackson, and others who followed in his literary footsteps, Bailey understood farming as foremost a moral enterprise, requiring farmers to learn about the laws of nature so they know how "to fit the crop-scheme to the climate and to the soil and the facilities." Again, in Bailey's words, "a man cannot be a good farmer unless he is a religious man."[9] It would be well to keep Bailey's words in mind as we consider over the longue durée the extent to which agriculturists, their supporting institutions, and consumers generally would adapt their activities and food preferences to protect Bailey's holy earth.

The first two decades of the twentieth century were prosperous times for farmers nationwide—in retrospect, so prosperous that in 1933 the farm lobby persuaded the US Congress to adopt the period August 1909–July 1914 as the base period for determining farm subsidies. With high food prices and

FIGURE 2.2. Liberty Hyde Bailey with plow. *Courtesy*, Division of Rare and Manuscript Collections, Cornell University Library, Ithaca, NY, Aggregated Photographs, #13-6-2957.

high profits, the number of Colorado farms more than doubled, to 59,934, between 1900 and 1920; the amount of farmland increased from 9.5 million acres to 24.4 million acres. Concurrently, the value of sales of farm products doubled and the value of farm property more than tripled during that period. Despite the fact that Colorado farmers produced a great variety of vegetables and fruits, their sales value in 1900 represented less than 3 percent of the total value of agricultural products compared to 50 percent for hay and forage crops and 20 percent each for wheat and sugar beets.[10]

The most dramatic increase in acres under cultivation took place on Colorado's eastern plains with the development of dry-land farming. To be clear, dry-land farming does not mean cultivation without water but rather cultivation without supplemental irrigation. Although new to Colorado, dry-land farming dates back to classical antiquity, popularized by the eighteenth-century Englishman Jethro Tull in his *Essay on the Principles of Tillage and*

FIGURE 2.3. Packing Colorado apples, 1900–20. *Courtesy*, Denver Public Library, Western History Collection, MCC 1900.

Vegetation, which emphasized deep plowing and frequent cultivation. Recall that the first settlers on the eastern plains had been attracted by unscrupulous or naive railroad agents during a period of unusually heavy precipitation; the agents told settlers that since "rain followed the plow," they could count on farming as they had done in humid eastern sections of the country. After drought returned in the late 1880s, many settlers abandoned their farms; some returned and left again in the mid-1890s. The few who remained and persevered would benefit from research at the USDA dry-land experiment station at Garden City in western Kansas and, beginning in 1908, the experiment station at Akron (Washington County) in cooperation with the experiment station at Colorado Agricultural College.[11]

Dry-land farming represents a remarkable effort to adapt to local conditions in an effort to enable the land to become productive. Much of what is known about agriculture on the eastern plains of Colorado at the turn

FIGURE 2.4. Dry-land pasture mix, Akron, Colorado, 1927. *Courtesy*, Agricultural and Natural Resources Archive, Colorado State University Libraries, Fort Collins.

of the twentieth century comes from published notes by James E. Payne, the Cheyenne Wells manager who had hired fellow Kansan Joshua Adams as a summer helper. Following the substation's closure, Payne spent three summers (1900–1902) as field agent for the college experiment station, documenting plant selection and cultivation techniques used by farmers and ranchers on the dry lands between the Arkansas and South Platte Valleys. Wheat was the first commodity crop planted by settlers, and Payne noted several varieties but was unable to clearly identify their origins. Farmers simply distinguished between wheat sowed in fall and spring, sometimes affixing their own family names to the varieties grown. For cultivating, conscientious farmers used two basic instruments: the plow with a double moldboard that moved soil to either side of a central furrow, followed by the harrow, a heavy frame with a smooth side dragged over the soil to cover seeds and a tined side to loosen soil and uproot weeds. The better farmers reported annual yields of 10 bushels per acre year after year in the same fields for up to ten seasons before noticing declines in yields due to deteriorating soil conditions. But even some of the best became discouraged by unpredictable, often extreme weather conditions and no longer prepared seed beds. To save on work, they took to broadcasting over uncultivated, often weed-infested fields, hoping nature would take care of the rest.[12]

Dispirited by their inability to count on constant wheat yields, dry-land farmers turned to diversification. They discovered that planting field corn in rotation improved wheat yields, which enabled them to use corn and wheat residues to feed hogs. Farmers also began to plant more drought- and wind-resistant grains such as sorghum and millet, varieties unknown. Although difficult to harvest because of its bulk, sorghum proved to be the all-around hardiest feed grain; millet best tolerated drought and heat, required the shortest growing season, and was easy to harvest. Some farmers tried growing alfalfa as a dry-land forage crop, although it required deep plowing, packing top soil, and planting just before well-soaked soil began to dry. Even then, an entire field of alfalfa could be quickly devoured by grasshoppers. Lacking insecticides, farmers discovered that by letting poultry into alfalfa fields, epidemics could be checked.[13]

In addition to crop failures in some years and low prices in others, grain farmers also had to face the difficulty of transporting their harvests to the railroad 30, 40, or more miles by wagon. "Considering these factors," Payne wrote, "almost any business farmer would decide to raise crops which could walk to market, or crops which could be condensed," by which he meant livestock and condensed milk.[14] All the more reason to protect the rangeland: with much still in the public domain and absent federal land management, growers had only their sense of self-interest in protecting forage grasses. Payne heard from old-timers that livestock losses due to harsh winters had declined as growers were keeping their stock closer to home, alive with the grains they themselves had cultivated. Despite such protection, losses remained high enough "to postpone the evil day indefinitely" when over-grazing would completely strip the range of nutritious grasses.[15]

At the Rocky Ford substation, Philo K. Blinn initiated breeding experiments to improve both yield and market value of the highly prized 'Netted Gem' cultivar of muskmelon (*Cucumis melo* L. var. *reticulatus*). Blinn's family had moved from Michigan to a farm near Berthoud in Longmont Colony, where Blinn grew up. As a student at Colorado Agricultural College, he worked on the campus experiment farm; after graduation, he transferred to the substation at Rocky Ford. Absent federal funding for substations, local growers contributed financially in support of Blinn's research. To be sure, melons had been grown in the Arkansas River Valley since well before his arrival. Grower J. W. Eastwood had obtained 'Netted Gem' seeds from the venerable W. Atlee Burpee Seed Company, having grown that cultivar on his Pennsylvania farm before moving west in 1885. Eastwood quickly realized that the dry climate combined with the availability of irrigation water was ideal for melons. When the first 'Netted Gem' melons were shipped from Rocky Ford in 1896, wholesalers considered their flavor so superior that for marketing purposes, they tied the melons to the locale, naming the variety 'Rocky Ford.' Although growers had experimented with improving its quality through seed selection, Blinn took a further step by seeking plants resistant to cantaloupe rust caused by a parasitic fungus (*Macrosporium cucumerinum* Ellis and Everh.). Experiment station researchers in Fort Collins had tested various fungicide sprays, including the first widely used commercial fungicide—a mixture of copper sulfate and lime developed by a professor of botany at the University of Bordeaux, France, thus the name Bordeaux mixture. The fungicide showed positive results, but the requirement of frequent spraying due to the rapid growth of cantaloupe vines made that approach cost-prohibitive. After two summers of trials on the substation farm, Blinn discovered that plants from the farm of J. P. Pollock were not only rust resistant but matured more quickly than other breeding lines, a significant factor since early cantaloupe fetched the highest market prices. An unusually altruistic farmer, Pollock welcomed Blinn's distribution of his seeds to other farmers.[16]

Seed selection alone was not enough to ensure best results. Blinn conducted painstaking cross-fertilization on the Pollock line over three generations to breed desired characteristics into a single specimen line. In doing so, he acknowledged applying the recently rediscovered and publicized principles of inheritance by the Augustinian monk and amateur naturalist Gregor Mendel. Blinn identified the 'Rocky Ford' melon as perfect or hermaphroditic, with male pollen-bearing organs (stamens) and female organs (pistils) residing in the same flower. He described how to carefully remove the unripe pollen about twenty-four hours before the flower opens, then place a paper sack carefully over the unopened bloom to protect the pistil from unwanted pollen; on the following day, manually apply the pollen to the stigma, the part of the pistil that receives the pollen during pollination. Since the State of Colorado as yet had no formal process to certify new cultivars and Blinn's improved seeds were traded from farmer to farmer, their integrity could easily be lost through cross-fertilization or careless handling.[17]

Philo Blinn understood that soil fertility and proper cultivation were just as important as seed selection; he also recognized that there was simply not enough barnyard manure available to keep up with the expansion of melon production. His predecessor at the Rocky Ford substation had conducted the first tests showing that melons grown in rotation with alfalfa matured earliest and with highest yields, which encouraged melon growers to supplement whatever barnyard manure they could find with alfalfa-melon rotation. During the 1904 cultivation season, growers experimented with inorganic (not derived from living matter) commercial fertilizers and discovered that even when they carefully followed manufacturers' directions, they lost plants when roots came into direct contact with those corrosive fertilizers. Blinn's advice to farmers: until we better understand the nature and proper application of those fertilizers, limit yourselves to barnyard manure and rotation with alfalfa. With the perfection of the 'Rocky Ford' breed, Arkansas Valley farmers extended their seed production from melons to include squashes, beans, peas, onion sets, seed potatoes, and other marketable vegetables. The combination of favorable growing temperatures, cool nights and warm days, low humidity to prevent disease, use of alfalfa, and the availability of irrigation contributed to making the Arkansas River Valley a major hub of seed production for the nation.[18]

At the same time the experiment station staff was working side by side with farmers, college faculty reached farmers in their communities by means of institutes or short courses, sometimes taught from railroad cars. As part of its public relations campaign to attract more freight business, the Denver and Rio Grande Western Railroad sponsored a "Potato Special" that traversed Western Slope counties and the San Luis Valley featuring lectures and exhibits. Faculty from agriculture-related fields such as chemistry, entomology, and engineering traveled to offer advice for addressing specific agricultural problems. As helpful as those activities might have been, the only farmers who attended were those who were truly interested, willing, and able to make the effort to participate. As is so often the case with educational outreach, absent were the farmers most likely to benefit. In an effort

FIGURE 2.5. Exhibit in railroad car, San Luis Valley, ca. 1915. *Courtesy*, Agricultural and Natural Resources Archive, Colorado State University Libraries, Fort Collins.

to overcome farmer reluctance, the agricultural college created the position of locally domiciled agricultural adviser, placing the first such advisers in the four counties of the San Luis Valley and in Logan County in 1912.[19]

The concept of assigning an agricultural adviser from an outside institution to a rural locale was not new. In France before the revolution of 1789, when the distinction between secular and ecclesiastical authority had yet to be clarified, the royal administration looked to recruit parish priests to promote scientific agriculture from their pulpits; these priests were often the only locally domiciled agents of centralized authority and the only literate inhabitants. Some priests supervised their own farms; others cultivated their own gardens. Of course, circumstances were entirely different in Smith County, Texas, where the position of county agent was first created in 1906. Its chief proponent was Seaman A. Knapp, former president of Iowa State Agricultural College who at the time was consulting with the USDA in Texas about eliminating the boll weevil from cotton fields. While a professor of agriculture at Iowa, Knapp had helped draft the Hatch Act (1887), which established and provided federal financial assistance to the network of agricultural experiment stations. Knapp was also credited with establishing the first publicly supported demonstration project on a farmer's private property.

Sharing with agriculturists the risks of new or unusual techniques would become the hallmark of the relationship among the federal government, the land-grant colleges, and the counties; it was institutionalized and made permanent by the Smith-Lever Act (1914), which authorized creation of the land-grant college-sponsored Cooperative Extension Service.[20]

As president of Colorado Agricultural College (1909–40), Charles A. Lory declared it his first priority to facilitate the transfer of useful knowledge to the state's farmers and rural communities—principally, application of the natural and physical sciences to the techniques of crop and livestock production. Lory had arrived in Colorado at age fifteen with his family, settling on an irrigated farm near Windsor; he spent two years as a ditch rider, patrolling irrigation works and water distributions to farmers. He earned a teaching degree at the state normal school in Greeley and a master's degree at the state university in Boulder and taught applied physics at the state agricultural college in Fort Collins. As president, Lory commissioned D. W. Frear of the experiment station and Daniel W. Working of the USDA, who was based on campus, to prepare and implement a plan to establish permanent positions of agricultural advisers in each of the state's counties. Among their earliest forays was successfully negotiating an agreement between the college and county commissioners in the San Luis Valley to hire L. M. Winsor, a graduate of Utah Agricultural College, to serve as permanent agricultural adviser under the auspices of the San Luis Valley Commercial Association. Frear, Working, and Winsor together hosted numerous listening sessions, some with individuals and others with groups, to ascertain how Winsor might find favor with reluctant farmers. Working was best attuned to the hopes and fears of agriculturists, having served as master of the Colorado State Grange and editor of *Colorado Farmer*. He would be Steinel's collaborator in preparing the 1926 history of Colorado agriculture.[21]

Winsor did gain farmer confidence by helping to address what they perceived as their most pressing issue: disease among hogs. He brought to the valley college veterinary professor George H. Glover, who diagnosed the disease as cholera. Glover convinced farmers to immediately destroy and dispose of all diseased hogs and immunize the remainder with an anti-cholera serum; over the long term, he strongly encouraged them to learn the principles and practices of farm sanitation. With no known cure for afflicted hogs, Glover noted that by sanitizing enclosures, farmers could slow the spread

FIGURE 2.6. Soil testing, 1928. *Courtesy*, Agricultural and Natural Resources Archive, Colorado State University Libraries, Fort Collins.

of the disease, lessening further losses. He persuaded hog farmers to establish their own improvement association, Monte Vista Hog Producers, and to continue to enlist extension expertise through their county adviser.[22] In 1912, Winsor reported that hog and sheep producers had added field peas to their repertory of forage crops. Of the same species (*Pisum sativum* L.) as garden peas but with seeds rounded rather than wrinkled, field peas represented a "conglomerate mixture" handed down through generations. Despite impurities, the mixture proved hardy and well-adapted to local conditions, requiring irrigation once before planting and no cultivation during growing season. To balance feed rations, the better farmers also grew alfalfa and barley, according to Winsor creating "a system of agriculture which cannot be surpassed." His report was unclear as to whether farmers fully recognized the advantage of the system in replenishing soil nutrients. Experiment station staff considered the system an excellent example of agricultural advisers and farmers working together for the sake of soil fertility.[23]

At the invitation of farmer George W. Seeley of La Jara, adviser Winsor helped establish and maintain a 5-acre cooperative demonstration project on

FIGURE 2.7. Crop improvement and standardization, barley, near Alamosa, Colorado, 1930. *Courtesy*, Agricultural and Natural Resources Archive, Colorado State University Libraries, Fort Collins.

Seeley's property. Winsor selected seeds, contributed his own manual labor, and kept records for approximately 100 crop varieties on five separate plots, drawing comparisons with other demonstration projects in the San Luis Valley. Principal crops included potatoes, peas, alfalfa, wheat, oats, barley, rye, rape, vetch, buckwheat, lentils, sugar beets, flax, corn, millet, sorghum, clovers, and a number of hay grasses. The two cooperators opened the demonstration project to inspection by any interested farmer, offered guided tours, and sponsored a Farmers' Day picnic luncheon followed by an afternoon of agricultural speakers from outside the valley.[24]

Adviser Winsor left the impression that his efforts were limited to crop and livestock improvement. His counterpart in Logan County conveyed a broader view of the agricultural adviser's role. D. C. Bascom was recognized as Colorado's first county agricultural agent. Appointed in 1912 by Colorado Agricultural College, his position was financed cooperatively by county and federal governments. True, Bascom's immediate task was to convince hesitant farmers that working side by side, he could help them improve their operations. Taking advantage of farmer interest in the Model T—by then, Ford had begun manufacturing conversion kits to turn the automobile into a tractor—Bascom offered tours of demonstration projects on surrounding farms. But Bascom also helped organize the testing of dairy herds, organized exchanges of farm equipment, promoted cooperative marketing, and campaigned for the establishment of tax-supported pest districts. With his background as a vocational agriculture teacher at Logan County High School, he sponsored youth farm clubs, popular ways to teach children about crops, ways to cultivate, and how to compete at county fairs.[25]

"Agriculture is not the whole of country life," President Theodore Roosevelt had written in his letter appointing Liberty Hyde Bailey to the Commission on Country Life in 1912. "The great rural interests are human interests, and good crops are of little value to the farmer unless they open the door to a good kind of life on the farm."[26] D. C. Bascom's approach to agriculture was in keeping with what would become official government policy. Two years after his appointment, the Smith-Lever Act institutionalized the partnership of county governments, the land-grant colleges, and the USDA by authorizing the establishment of the Cooperative Extension Service. Its mission: "to aid in diffusing among the people of the United States useful and practical information on subjects relating to agriculture and home economics, and to encourage the application of the same." By specifying home economics, the act sought to address the fundamental question posed by the commission: "How can the life of the farm family be made less solitary, fuller of opportunity, freer from drudgery, more comfortable, happier, and more attractive?" In some quarters, that broader mandate remains a subject of controversy, yet it gave the extension service the flexibility to adapt its programs to the coming decline in the farm population and the gentrifying of country living.[27]

At the time of the US entry into World War I, county agents were the only federally subsidized workers (except for postal employees) in most rural areas. As a result, the federal government turned to the extension service to administer wartime agricultural policies. Emergency appropriations authorized by the Food Production Act of 1917 enabled Colorado extension to place agents in most counties. A companion piece of federal legislation, the Food and Fuel Control Act of 1917, established the US Food Administration and charged the food administrator with controlling all aspects of food production and distribution: increasing production to feed the military as well as civilians in war-ravaged Europe while decreasing consumption and keeping food prices low at home.

In addition to serving as county-level federal bureaucrats, the county agents were pressed by the government to push sometimes reluctant farmers to increase production of commodity crops, especially wheat. By fixing the price of wheat at a relatively high level and establishing a grain corporation to buy and sell wheat, the federal government inaugurated that delicate and often controversial policy of balancing supply and demand, seeking to ensure fairness for all concerned. In the name of increasing production, county agents tended to favor larger businesslike producers over smaller, more tradition-bound farmers, using their administrative authority to funnel seed, fertilizer, machinery, laborers, and agricultural credit. To support the county agents in their endeavors, President Lory called together a special committee of faculty, experiment station staff, and extension specialists to coordinate on-campus efforts on food production and food conservation. Because of wartime mobilization, soil and water conservation principles and practices taught at Colorado Agricultural College and transmitted by at least some county agents were set aside for more than a decade. That was especially noticeable on the eastern plains, where high wheat prices combined with favorable moisture conditions attracted legitimate investors and new farmers but also speculators known as "suitcase farmers." The latter caused many ills, according to the Cheyenne County agent who, long after the speculators had disappeared, would be assisting farmers in dealing with soil erosion, the worst of those ills. In their rush to benefit from increased production, farmers had purchased on credit great numbers of tractors and other equipment, working day and night to convert all accessible virgin grasslands to wheat.[28]

Among the obstacles preventing Colorado wheat growers from ramping up production and competing on the open market was their use of seed mixtures. Prior to 1917, Colorado had no pure seed law; farmers had obtained seeds wherever they could, often from their own fields. There had been a few isolated efforts to grow pure lines; even after 1917, though, some farmers continued to show little interest in varietal purity. For wheat growers throughout the Great Plains, a breakthrough came in the development and distribution of the cultivar known as 'Kanred.' As its contracted name suggests, 'Kanred' was a hard red winter wheat cultivar developed at the Kansas

Agricultural Experiment Station. Its origins can be traced back to a Crimean type, introduced into the United States by the USDA in 1900 and distributed to experiment stations for research purposes. With heads of 554 specimens, economic botanist Herbert F. Roberts created the actual selection known then as 'P-762.' In 1910, Kansas Agricultural Experiment Station agronomists started testing the selection, first in nursery rows and then in field plats to obtain more accurate yield records. Further trials took place at the branch station near Hays and on cooperative demonstration plats throughout Kansas. By fall 1917, the Hays station had amassed enough foundation seed to distribute 'Kanred' for commercial production. Colorado could initially obtain seeds from county agents in Boulder, Morgan, El Paso, and Weld Counties and within three years from agents in eleven additional counties.[29]

By 1920, 'Kanred' had become the cultivar most recommended by Colorado county agents. Extension agronomist A. E. McClymonds noted that compared to 'Turkey' and 'Kharkof,' the two most commonly planted types of winter wheat, 'Kanred' best tolerated winter cold; produced more stems from its roots, meaning fewer seeds had to be planted; ripened slightly earlier; proved more resistant to rust; and, most important, yielded about 4 bushels per acre more than the other cultivars. During the 1917–20 seasons, 'Kanred' yielded an average of 22.3 bushels per acre at the Akron experiment station compared to 15.8 bushels for 'Kharkof.'[30]

For wheat farmers opposed to price setting by the federal government, field corn became a profitable alternative. Recall that field corn brought to Colorado from the East produced far larger cobs than "Mexican" corn. As early as the 1880s, Ainsworth Blount had tested several varieties of field corn on the college farm; individual farmers, too, had sought to hasten the process of improvement through selection. Sometime in 1915, Boulder County agent H. H. Simpson organized a corn-improvement project for the boys' corn clubs under his supervision. He chose to use seeds of 'Minnesota No. 13,' a cultivar already successfully dry-land farmed by at least one farmer on the eastern plains. Developed at the Minnesota Agricultural Experiment Station for adaptability to the short growing season in the upper Midwest, 'Minnesota No. 13' offered the additional advantage for Colorado of low-growing stalks, making it less prone to wind damage. The cultivar belonged to the dent type of corn—yellow kernels indented at their tips when mature, 12 to 18 kernels per row, and single reddish-brown cobs 7 to 8 inches in length. As an open pollinated type—pollinated by insects or wind—'Minnesota No. 13' seeds saved from one year's crop would produce exactly the same 'Minnesota No. 13' the next year and every year thereafter. Having selected the seed, Simpson went into the fields with the youths, teaching about soil preparation, planting time, seed spacing, cultivation, and seed storage. He required the youths to keep careful records of germination rates with or without irrigation and, most important, of crop yields. As he had hoped, corn club successes encouraged commercial farmers in the county to try 'Minnesota No. 13.' For those farmers, Simpson had secured "loans" of seeds directly from the Minnesota experiment station; for every bushel of seeds received, farmers had to return 1 bushel from their own harvests. Simpson noted that by adopting the cultivation practices demonstrated by the youths, even the farmers who declined to adopt 'Minnesota No. 13' still gained in yields and profits. By 1920, 'Minnesota No. 13' had become the standard corn cultivar for the county.[31]

Unquestionably, the wartime push to increase food supplies added a sense of urgency to the production, collection, and distribution of clean, high-quality seeds. With the likely cut in seed shipments from overseas, major eastern seed houses turned to Arkansas Valley farmers to supply seeds to their specifications. In April 1917, the legislature passed and Governor Julius C. Gunter (D) signed the Colorado Pure Seed Law. It required that all field seeds sold in Colorado be tested and labeled by the state seed-testing laboratory at Colorado Agricultural College. Labeling did not guarantee seed purity, only that the contents of a seed sack were accurately described on its label—listing the commonly accepted cultivar name, where the seed was grown, pure seeds as a percentage of total sack content, rate of germination, date of testing, and the presence of weed seeds by common name and by number if in excess of ninety seeds per pound. By failing to fund inspection and enforcement, the seed law did not stop the continuing sales of impure seeds. The law was repealed, reenacted, amended, and re-adopted several times, most recently in 2017 as the Colorado Seed Act, which replicated the federally recommended uniform state seed law.[32]

Conscientious farmers always understood that better seeds produced higher yields. With greater emphasis on seed research at the experiment station, farmers came to rely on commercial seed houses offering seeds that produced plants "pure and true to type," that is, free of extraneous matter and guaranteed to yield in every future generation plants with the

FIGURE 2.8. Registered seed exhibit, Boulder County, 1922. *Courtesy*, Agricultural and Natural Resources Archive, Colorado State University Libraries, Fort Collins.

same characteristics as the original plants. In 1918, extension agronomist A. E. McClymonds began drafting standards for the production of certified seed that prospective seed growers could easily follow. Beginning with field corn, he suggested that certification be made on a county-by-county basis by a committee of three under the direction of the county agent. In Logan County, agent J. E. Morrison reported the following certification standards: "No mixture of variety, maximum 0.5% other crop, recleaned to 98% pure, no noxious weeds, 0.5% maximum of other weeds, germination 90%." As interest in certified field crops grew, the state department of agriculture created the position of state seed certification and registration agent. As the first occupant of that position, McClymonds developed a set of uniform standards and practices for certifying corn, wheat, barley, rye, alfalfa, and soybeans. In 1922 he helped the Colorado Seed Growers Association (est. 1918) organize the first state-certified seed show, which took place in Colorado Springs with exhibits from five counties.[33]

In addition to protecting legitimate sellers and buyers, the pure seed law was meant to reduce the introduction and spread of weeds, whether invasive or native. Subject to the ninety seeds exception, the law prohibited the sale of sacks containing one or more of nine primary noxious weeds, most notably

FIGURE 2.9. Russian thistle blown into ditches near LaPorte, 1920. *Courtesy*, Agricultural and Natural Resources Archive, Colorado State University Libraries, Fort Collins.

bindweed (*Convolvulus arvensis* L.), Canada thistle (*Cirsium arvense* [L.] Scop.), and leafy spurge (*Euphorbia esula* L.); it also required that twelve secondary noxious weeds be listed on labels. At the time, Canada thistle posed the most serious threat, propagating through underground stems; a small patch in one part of a field could spread underground to another part or even over the entire field. To be sure, one farmer's weed may not be another's; a weed's noxiousness depends on the extent that it interferes with the cultivation of plants or in some cases the production of livestock.[34]

In 1919, experiment station botanist Wilfred W. Robbins and agronomist Breeze Boyack published the first non-technical guide to help Colorado farmers and ranchers identify and control weeds. Citing USDA statistics on the economic losses in yields and labor, they argued that weed control required intelligent cultivation and county support through furnishing supplies and machinery. They stressed the use of pure seeds, well-composted manure (without weed seeds), crop rotation, alfalfa being best for crowding

FIGURE 2.10. Spraying herbicides, San Luis Valley, 1916. *Courtesy*, Agricultural and Natural Resources Archive, Colorado State University Libraries, Fort Collins.

out weeds, and cutting weeds around fields; for longer-term control, they recommended cutting along roadsides and ditches and on vacant farmland. In addition, they urged caution with the use of herbicides. Experiments had shown that some chemicals effectively destroyed broad-leaf but not narrow-leaf plants, which did allow for application on grain fields. Available chemicals were common salt, copper sulfate, iron sulfate (copperas), sodium arsenate, calcium chloride, kerosene, carbolic acid, and sulphuric acid. Iron sulfate appeared to be the chemical that was most effective, least expensive, safest to handle, and not poisonous to livestock. Overall, however, the preferred means of weed control remained careful cultivation and crop rotation. Natural chemicals such as sulphur had been used by agriculturalists since Homeric times; synthetic chemicals were yet to be invented.[35]

By controlling weeds on roadsides, fencerows, and ditch banks, Robbins and Boyack noted that farmers could obtain another benefit: less natural habitat for insect pests to deposit their eggs and thus fewer hatched larvae to devour crops. On the other hand, without weeds surrounding cultivated fields, the pest population would be forced to settle on cultivated plants—another reminder about the imprecision of the term *weed* and the fact that not all weeds are noxious, which would pose real dilemmas for managers of those special tax districts established to underwrite the costs of

FIGURE 2.11. Grasshopper control using Paris green, San Luis Valley, 1916. *Courtesy*, Agricultural and Natural Resources Archive, Colorado State University Libraries, Fort Collins.

pest control.[36] Among insect pests, grasshoppers arguably caused the most damage. Recall that to control the potato beetle, Greeley farmers had used the arsenic poison Paris green. With the advent of spraying by mechanical means, Paris green became the insecticide of choice, although researchers at the experiment station still favored the use of less drastic alternatives.

Entomologist Charles R. Jones distinguished between natural and artificial pest control: the former included predaceous and parasitic insects, as well as birds; the latter, cultivation, poison sprays, poisoned baits, and a variety of mechanical means to eliminate grasshoppers. He reported on breeding experiments with *Sarcophaga*, a flesh-eating fly first noted feeding on grasshoppers around freshly stacked hay near Trinidad in 1916. But by far the most effective natural control came from birds, as field agent James Payne had reported as early as 1901. In a farmstead near Wray (Yuma County), he had seen a vegetable garden entirely devoured and fruit trees entirely stripped of leaves; two years later, he found both garden and trees in good condition. The farmer explained that he had brought in 400 chickens and turkeys to control grasshoppers, with turkeys favored as wanderers over chickens, which needed to be moved around by the farmer in affected areas. Entomologist Jones cautioned, however, that fowl can effectively control only small areas; no farmer could keep enough fowl to cover entire fields. Clean cultivation along ditches and fencerows remained the preferred method for preventing epidemic outbreaks, but it required cooperation among farmers and the financial support of county commissions and state government.[37]

The emergence of two leading farm organizations during the first two decades of the twentieth century helped bring farmers together, although the groups' respective notions of cooperation, as well as their ideals, policies, and practices, quickly diverged. First in time, the Colorado Farmers Union envisioned a future of family farms, with an emphasis on economic and social justice for all, not just farmers; the Colorado Farm Bureau, first in membership, envisioned a future of profitable farms regardless of size, with an emphasis on profit like any other business. The farm bureau would become a leading advocate for mechanization, standardization, bigger farms, and those efficiencies that produced higher yields and greater profits—which closely paralleled USDA policy. The farmers union remained committed to a more traditional notion of farming as the most ennobling profession and of farmers as the bedrock of the republic; in time, it was more inclined to support conservation of land and water as an intrinsic value and was critical of USDA policy that favored "bigger is better."

Just as the farmers' alliances overtook the Grange, the farmers union succeeded the farmers' alliance. The farmers union was formed in 1902 in the cotton country of northeast Texas by nine men—six dirt farmers, a schoolteacher, a county clerk, and a country doctor—brought together by a former organizer for the farmers' alliances who convinced the group to hire him as membership recruiter. Like the Grange, the farmers union began as a secret society, spread rapidly, overcame startup squabbling, and established itself as the Farmers Cooperative and Educational Union of America, with each state division having an equal voice in the national organization.[38]

The Colorado division of the farmers union began in 1907, founded in a most unlikely place—a cabin in Crystola in the mountains west of Colorado Springs—by a tiny group of religious and political eccentrics convinced that farmers, like industrial workers and urban laborers, had been dispossessed by unfettered capitalism. At their constitutional convention in Pueblo, they set forth a broad moralistic outlook: to establish justice, secure equity, and apply the golden rule. Underlying their program of action was a conviction that cooperation through buying and selling cooperatives was the best vehicle to improve the lot of farmers and ranchers and preserve and strengthen their communities. In that way, founders of the farmers union shared the same faith in progress expressed by the founders of the Greeley and Longmont colonies. Not to be overlooked, James G. Patton (1902–85), the most illustrious leader of the Colorado Farmers Union and later the National Farmers Union, grew up in a little-known colony in southwestern Colorado: the New Utopia Cooperative Land Association (Nucla). During his tenure, he borrowed ideas and strategies and hired key staff from the more radical, short-lived farm organizations such as the Farmers' Alliance, Nonpartisan League, and Farm Holiday Association. Of all the groups operating in Colorado, only the farmers union would remain true to its populist roots and progressive ideas, surviving and even thriving to the present day.[39]

The American Farm Bureau Federation was not organized until 1919, but independent farm bureaus had begun in 1911 for the purpose of supporting agricultural business agents who offered not only advice to dues-paying individual farmers but purchasing and marketing services as well; those agents often worked from desks in local chamber of commerce offices, thus the name "farm bureau." With passage of the Smith-Lever Act establishing county agents as extension service employees, farm bureau agents could no longer enter into direct business relationships with individual farmers, although county agents continued to devote substantial portions of their time to farm bureau member recruitment.

In 1916 the Cooperative Extension Service launched a plan for the creation of county-wide farm bureau auxiliaries meant to promote and support the educational activities of county agents. Initially, farmers correctly viewed the extension service and the farm bureau as synonymous. In Peetz (Logan County), for example, the local farmers union secretary referred to the "Colorado Farm Bureau Extension" when announcing that the local would sponsor the county agent and home demonstration agent for a four-day educational program. Given wartime mobilization, whether at Peetz or elsewhere, county agents pressed for greater production of crops; the home demonstration agents advised on food preservation, especially canning vegetables and fruits, and suggested substitutes for wheat and other commodities in short supply domestically.[40]

By 1919, recruitment to farm bureau county auxiliaries had advanced sufficiently to warrant the formation of an autonomous statewide organization separate from the extension service. Holding its organizational meeting on the campus of Colorado Agricultural College, the Colorado Farm Bureau Federation began with nineteen county chapters. Its stated purpose: "to develop, strengthen, and correlate the work of the county farm bureaus in their efforts to promote the development of the most profitable and permanent system of agriculture suitable for Colorado's conditions; the most wholesome and satisfactory living conditions; the highest ideals in home and community life."[41] As farm bureaus nationwide grew in popularity, they developed into more than mere booster clubs for county agents. Because their precursors, the farm bureau agents, were in reality business developers, the business community turned to the farm bureaus as bulwarks against farm and labor organizations that advocated systemic economic and social change. Under the cloak of patriotism, farm bureau leaders in the Midwest drew a direct connection between the Russian Revolution of 1917 and an array of American organizations they considered dangerously socialistic. Taking advantage of the "Red Scare," those leaders had formed the American Farm Bureau Federation.

The farm bureaus and their business supporters most feared the Nonpartisan League, which had its ideological roots in the Grange, the farmers' alliances, and the Populist Party. The Nonpartisan League advocated what it called economic democracy, described by historian Michael J. Lansing as "a moral economy premised on accumulation without concentration."[42] In its political strategy, however, the league differed from the other groups by supporting candidates pledged to support its program, especially effective in the recently inaugurated system of party primaries. The league was most active in North Dakota, electing its candidate for governor and fielding a majority of state legislators who voted to establish the state-owned Bank of North Dakota and the state-owned North Dakota Mill and Elevator Company. Spreading out from North Dakota, in its heyday the league claimed over 250,000 members, primarily in grain-producing states including Colorado.[43]

In early 1917, the Nonpartisan League sent Ray McKaig, master of the North Dakota Grange, to speak on the league's behalf to the Colorado Federation of Farm Organizations meeting on the Colorado Agricultural College campus, to delegates at the annual state Grange convention, and to labor

FIGURE 2.12. Grain elevator, Craig, 1915. *Courtesy*, Denver Public Library, Western History Collection, MCC2554.

leaders in Denver. He also met separately with the farmers union, whose goals were most compatible with those of the league. Indeed, farmers union president James M. Collins, a livestock and grain grower from Eaton, had taken the initiative to bring together both statewide farm organizations and emerging commodity groups—most of them started with assistance from the farmers union—in a united effort to lobby the state legislature.[44]

As Colorado was becoming more urbanized, Collins pushed for reapportionment of the state legislature to give rural voters a greater voice. This reflected a farmers union principle that as stewards of the land and water, agriculturists had a special responsibility to engage in the affairs of government to protect those resources. By the time of the 1918 election, the Nonpartisan League had secured a hold on the Colorado Democratic Party; with the support of other farm and labor groups, it selected Collins as the Democratic candidate for governor. Failing to carry the cities, Collins lost the election to Republican Oliver H. Shoup. Despite opposition by business interests accusing leaguers of being Bolshevik sympathizers, the league did manage to elect four partisans to the state legislature. The league claimed 12,000 members in December 1919 and then slowly withered as a result of internal strife, leaving the farmers union as the progressive voice.[45]

With the advantage of hindsight, one can fairly state that from the beginning, the American Farm Bureau Federation and its state affiliates have been politically right-of-center, generally favoring Republican policies and candidates; the National Farmers Union and its affiliates have been left-of-center, generally supporting Democratic policies and candidates. Both organizations favor some form of federal support of agriculture, agreeing in general that agriculture requires government engagement in achieving and maintaining parity between farmers and those in urban and industrial occupations. The farm bureau advocates for price supports and federally backed crop insurance, leaving the rest to market self-regulation. The farmers union wants a federal guarantee of farm income. But parity was not the only balance required; equally essential, policymakers and the public realized belatedly that the federal government needed to set the standards and provide the incentives for farmers and ranchers to conserve land and water to ensure a continual supply of food and fiber for the nation.

3

Stretching Nature's Limits

In the wake of wartime mobilization, farmers and ranchers needed to produce more, which meant purchasing more land, using more machinery, and adopting innovations in the agricultural sciences that contributed to higher yields. But as is frequently the case with discoveries, not all of the long-term effects were known at the time. Perhaps most notable were the cases of pure seeds, natural chemicals, and manufactured fertilizers. That is not to deny the very real benefits of discoveries. It is simply that agriculturists were pushing against nature's limits, not yet recognizing that such limits existed or if they existed, where those limits might be. In conveying discoveries from laboratory to farm and ranch, county agents were in the unique position of advising caution in their application, trying to make sure proven techniques would remain in place.

Visitors to county fairs in fall 1926 would have seen a scale model of an 80-acre farm, part of a larger extension service exhibit to demonstrate that diversified farming provided higher yields while maintaining and even improving soil fertility. Homesteaded in the early 1870s, the actual farm, located

https://doi.org/10.5876/9781646422050.c003

near Greeley, had experienced six different owners, at the outset growing corn and grain as cash crops and enough hay to feed a few dairy cows and work horses; with improved irrigation, the farm added potatoes and then sugar beets after a processing plant was built nearby. Sometime around 1910 the farm owner began to feed livestock, having shifted one-quarter of the property plus an additional 80 acres leased to cultivate alfalfa. Even with the decrease in crop acreage, crop yields improved because of increased application of manures and adoption of a four-year rotation cycle. Dividing the farm into four fields, the farmer successively cultivated each field with alfalfa, potatoes, sugar beets, and small grains. To conserve soil fertility, the farmer plowed each field only once every four years: plowing under alfalfa for potatoes, using a spring-tooth harrow to prepare for planting beets, and then seeding small grains to alfalfa. Growing several varieties at once lessened the impact of a single crop failure, which the farmer further reduced, when time permitted, by replanting an overworked field with dry beans. By adding the livestock-feeding component, the farmer could afford to hire labor year-round, an attractive consideration for retaining good workers.[1]

Intentionally or not, in describing a successful diversified farm operation, the extension agents presented a practical as well as a conservative way for a farmer to survive periods of depressed prices. Recall that to meet wartime demands, the federal government had pushed farmers to produce more. As European agriculture recovered, demand for American farm products declined; prices paid to farmers dropped, but their costs did not. To cover rising costs, farmers generally sought more efficiency and bought more land, which contributed to higher debt and made their long-term financial conditions even more tenuous.[2] To be sure, wartime mechanization had helped farmers overcome labor shortages; postwar mechanization saved on labor costs, although only on larger farms where tractors and other motorized machinery made economic sense. Both the farm bureau and the farmers union were essentially correct in their predictions about the future of American agriculture: mechanization would mean that larger farms would run like any other big businesses and that there would be fewer smaller farms and communities that depended on them. During the 1920s and early 1930s, the extension service, the experiment station, and the US Department of Agriculture (USDA) advocated mechanization regardless of the size of farm operations.

FIGURE 3.1. Soil sampling, 1930. *Courtesy*, Agricultural and Natural Resources Archive, Colorado State University Libraries, Fort Collins.

Despite their efforts, extension agents reported again and again on the persistence of apathy and ignorance among farmers, who continued to rely on past practices rather than trying new techniques proven to increase yields. In Conejos County, agent Frank R. Lamb documented years of declining yields, which he attributed to degradation of soil fertility caused by poor cultivation practices: continuous cropping to grains, allowing hogs to feed on peas in the same fields year after year without plowing, over-irrigating, shallow plowing, and inadequate manuring. Lamb contended that the extension service needed to do more to educate farmers, starting with the "simple practical methods of improving the physical condition of the soil before attempting any work of a scientific nature." The most effective way to teach reluctant farmers was by establishing more demonstration projects on individual farms and having agents work shoulder to shoulder with them.[3]

Through demonstration projects, county agents promoted practices that, for the most part, were rather traditional—seed selection, crop rotation, manuring, and cultivation to fit local conditions. Their first piece of advice for improving yields concerned the use of pure seed. The reality that most farmers needed to make greater efforts in securing true-to-type pure seeds could be observed at any grain elevator, according to extension agronomist Waldo Kidder: "As one watches the loads of grain being dumped by the farmer, he learns that each of the growers is taking a dock of 5–7 cents [per hundredweight] because the grain is mixed or has too much weed seed in it or from other preventable causes." But if farmers plant pure seed on clean ground, in the rotation cycle most appropriate to their specific soils, they will harvest several more bushels per acre. As pure seeds first became commercially available, there were few convenient rural locations where farmers could purchase those seeds. County agents had generally failed to convince farmers that the transportation costs of obtaining seed over long distances would be more than offset by higher yields and bigger profits. Farmers continued to use seeds left over from prior years or purchased from neighbor farmers or from an occasional, often unscrupulous traveling salesperson.[4]

Agronomist Kidder led an effort to work out a system in which experiment station staff or trusted growers supplied certified seed to the farmers. The experiment station secured breeder seed, that is, seed directly from the source of creation—for example, the popular 'Kanred' cultivar of wheat from Kansas Agricultural College. Agronomists then evaluated the breeder seed for quality and suitability to local growing conditions; if satisfied, they then distributed it as foundation seed for Colorado seed growers to produce for commercial distribution. After harvest, seed growers sent samples to the experiment station for bin inspection and then to the seed laboratory for purity and germination tests. If the seeds tested 92 percent pure and high in germination, the agronomy department assigned a registry number to the sample. As long as the selected grower produced seeds equal in quality to the sample stored with the agronomy department, the harvested seeds would be considered pure. To assure buyers that a given variety retained its identity, the selected farmer signed a certificate of authenticity attached to every lot of certified seed up for sale. To this day, we refer to the four phases of a seed's development as breeder, foundation, registered, and certified seed.[5]

After passage of the Colorado Pure Seed Act in 1917, agronomists and extension staff sought recommendations from certified growers in setting specific rules and regulations and determining means to distribute certified seeds. In return, staff expected growers to keep complete and accurate records, sharing their cultivation practices and results with the extension service. By fall 1925, extension agronomist Kidder and county agents had succeeded in recruiting 160 certified seed growers to form county crop improvement associations. The county groups made up the membership of the revitalized Colorado Seed Growers Association, which elected Kidder as its first permanent secretary. Extension agents relied on association members to convince fellow farmers about the advantages of exclusively using pure seed and becoming more competitive with growers in states ahead of Colorado in the use of certified seed, all the while reducing grower expenses for controlling weeds and plant diseases. Operating under the auspices of the extension service, the Colorado Seed Growers Association continues to serve as the state's official seed certifying agency.[6]

Intentionally or otherwise, increased use of certified seed contributed to reductions in the variety of crops grown, discouraged farmers from experimenting with less common varieties, and contributed to the cultivation of single crops to the exclusion of others. The extension service encouraged that tendency through its sponsorship of yield clubs that encouraged youths to grow standardized crop varieties; those clubs eventually melded into the 4-H organization, also sponsored by extension. The long-term impact of monoculture on plant diversity, soil fertility, and food security was yet to be thoroughly understood.[7]

Certification did extend to potatoes that were grown as seed potatoes. Certified growers set out potatoes—small potatoes whole and larger potatoes cut up so that each piece had two or three eyes (root buds) and a small

FIGURE 3.2. Poster, 1931. *Courtesy*, Agricultural and Natural Resources Archive, Colorado State University Libraries, Fort Collins.

amount of surrounding tuber. County agents generally agreed that more than any other factor, the use of certified seed potatoes contributed to the success of the potato as a commercial crop. Perhaps that was because

FIGURE 3.3. Irrigating potatoes, 1900–1920. *Courtesy*, Denver Public Library, Western History Collection, MCC 1847.

certified potatoes were less affected by the two main potato diseases: the fungus *Rhizoctonia solani*, most prevalent in irrigated fields, and the fusarium fungus naturally present in most soils. Following crop failures in the Greeley area and pleas by area growers for assistance in finding remedial or preventive measures, the USDA Bureau of Plant Industry established a potato experiment station northeast of Greeley in 1915, transferring supervision to Colorado Agricultural College in 1924. In explaining the presence of potato diseases, resident station plant pathologist Howard S. MacMillan noted that plants in their native places were generally disease-free or at least able to survive disease, but "the irrigated west offers as artificial an environment as the potato encounters anywhere throughout the country." To be sure, yields are large, conditions for growth are favorable, and risks are minimal during the growing season, "but for all of this the irrigated potato is in a strange land." Acknowledging that disease can never be entirely prevented,

MacMillan recommended that the grower use only "good seed of the right variety for his soil," preparing the field well, planting properly, and exercising "genius [imagination and intuition] in cultivating and irrigating."[8]

Not unlike the practice of medicine, farming is both an art and a science, which may help explain why Colorado Agricultural College horticulture professor Emil Peter Sandsten expressed the same sense of frustration toward farmer apathy and ignorance with regard to soil fertility that had been reported time and again by county agents. Sandsten wanted to attribute that way of thinking to the relative newness of farming in Colorado, with soils deteriorating slowly over time: "Yet every grower should realize that the process of depletion is much more rapid than accumulation and that it is not too soon to take measures to maintain the fertility of our soils."[9] Arriving on campus in 1913 from the University of Wisconsin, Sandsten was arguably the best-trained and most experienced faculty member at Colorado Agricultural College to date. Born in Sweden, he had immigrated to the United States at age eighteen and worked as a market gardener before entering the University of Minnesota for bachelor's and master's degrees in horticulture. Prior to earning a PhD in horticulture at Cornell University and starting his teaching career, Sandsten had spent at least one summer in Colorado working in agriculture under irrigation. In 1914, Governor Elias M. Ammons (D) appointed Sandsten as the first state horticulturist.[10]

Although recognized primarily for his work on orchard fruit and small fruits, Sandsten necessarily dealt with market produce and crops, notably potatoes. Between 1920 and 1930, the state's potato acreage increased by 14 percent and production by 60 percent, making potatoes Colorado's highest-valued crop.[11] Most potato growers were inclined to believe that using alfalfa in rotation, harvesting the first cutting for livestock feed, and turning under the second cutting sufficed to conserve soil fertility indefinitely. They understood that alfalfa supplied both plant matter (humus) and nitrogen through the nitrogen-fixing bacteria that cling to alfalfa roots. Sandsten agreed on the benefit of alfalfa. But he held the view that to produce maximum yields, potatoes needed the addition of potash and phosphorus, supplied most economically by barnyard manure. He foresaw the day when potato growers would have to rely on commercial fertilizers.[12]

Potato growers, meanwhile, had started using sugar beets in their rotations or shifting to sugar beets alone. Beets required substantial acreages

FIGURE 3.4. *Left to right*: Governor Edwin C. Johnson, Dr. Sandsten, and T. G. Stewart, wheat field, 1936. *Courtesy*, Agricultural and Natural Resources Archive, Colorado State University Libraries, Fort Collins.

to realize profits, in addition to the large capital investment in refineries far beyond the means of individual growers. By the 1920s, the Great Western Sugar Company owned and operated processing plants in twelve northern Colorado communities; the Holly Sugar Company owned plants in Delta, Grand Junction, and Swink near Rocky Ford; and the American Beet Sugar Company owned a plant at Rocky Ford.[13] Just as the railroads sought to build their businesses by recruiting farmers to settle new lands, the sugar companies found it in their interest to help farmers raise the sugar content of their beets. In 1910 the Great Western Sugar Company established its own experiment station at Longmont where scientists and technicians conducted research on breeding, cultivation, weed and pest control, and the development of specialized machinery for planting, cultivating, and harvesting. Much like county agents, company field representatives conveyed to growers what was learned in the laboratory and helped farmers apply the results to their farms. Research at the Longmont station confirmed the USDA's own research, which showed that growing sugar beets in rotation increased yields of all crops. As a result, Great Western field representatives became outspoken advocates for diversified agriculture.[14]

During January and February 1927, the extension service in cooperation with two refiners—Holly and American Beet Sugar—and the agricultural marketing department of the Santa Fe Railroad planned an exhibit train through the Arkansas Valley. Extension agronomist Waldo Kidder and his colleagues took up an entire baggage car to demonstrate the benefits of crop rotation. Upon entering the car, visitors saw a revolving wheel divided into eight segments, under the banner "Crop Rotation Pays, Put Your Faith on This Wheel of Fortune for the Arkansas Valley." Mounted on each segment was a crop species that could be planted as part of the recommended eight-year rotation cycle. The exhibit included a series of illustrated panels depicting in detail how the various crops could be used. Below the panels, the extension staff had built a model of an actual Arkansas Valley farm for which records had been kept for the prior five years. Kidder described its rotation as follows: small grain field seeded to alfalfa in year one; alfalfa harvested for three years, with the last cutting plowed under while still green to help restore soil fertility; field planted in spring to corn, small grain, or vine crops; next year field manured and planted to sugar beets; following year, to small grain and sweet clover, with sweet clover plowed under the same fall or early the next spring; field planted to sugar beets, vine crops, or corn, completing the cycle. Kidder noted the conservation benefit of including sugar beets in rotation. Like alfalfa, beet taproots grow vertically down far enough to avoid the cultivator's equipment—leaving root residuals, helping to break up the soil, allowing for aeration, letting water seep through the soil, and retaining more moisture. All in all, the actual farmer had "made good money, so we, in miniature, showed his farm which excited a good many comments and a great deal of interest."[15]

Dispersing practical information to farmers at no cost helped those receptive to best practices; offering premium prices for beets with high sugar content provided incentive for everyone. When the Colorado sugar beet industry began in 1900, 1 acre planted to beets produced 1,000 pounds of granulated sugar; by 1930 that same acreage produced 6,000 pounds. In addition, both growers and refiners had learned to make efficient use of beet tops and pulp, resulting in by-products—mostly in the form of molasses—that, when mixed with alfalfa and grain, helped fatten livestock. In 1930 more than 1 million sheep and lambs in northern Colorado feedlots were fed on beet by-products.[16]

The sugar beet industry did not, however, depend on scientific agriculture alone for its well-being during difficult times. The US Congress did protect domestic sugar production; making matters more politically complex, however, was the reality that it was more expensive to produce sugar from beets than from cane. Still, production in the northern states exceeded the more established cane production in the southern states.[17] Moreover, the capital-intensive nature of beet processing—combined with the absence of competition among refiners and the fact that refiners set the standards for planting, cultivating, and harvesting—convinced growers that they operated at the mercy of monopolists. Federal investigations confirmed their opinion. To make matters worse, by keeping payments to growers low, refiners had forced growers and their laborers to put their children in the fields, keeping them out of school for weeks in the spring to help with cultivation and for months in the fall for the harvest. Add to that the postwar decline in farm prices and one can understand why growers sought to organize themselves into commodity associations.[18]

During the early 1920s, beet growers, wheat growers, and other commodity groups banded together with the American Farm Bureau Federation and the National Farmers Union in support of the newly formed farm bloc, a bipartisan group of congressional members pushing for legislation favorable to farmers. The farm bloc's earliest accomplishment was passage of the Capper-Volstead Act in 1922, hailed by both the farm bureau and the farmers union as the great charter of the cooperative movement. The act authorized agricultural producers to associate together in marketing their products without fear of anti-trust prosecution. Lobbied most heavily by the state chapter of the farmers union, the Colorado legislature passed companion legislation in 1924. The intent of the state act was to promote orderly marketing of agricultural products through cooperation, eliminate speculation, encourage the most direct possible distribution of agricultural products from producers to consumers, standardize the marketing of agricultural products, and provide for the organization and recognition of cooperative marketing associations. The Mountain States Beet Growers' Association was among the early beneficiaries of the act, serving as the exclusive agent for a majority of growers in negotiating with the refiners. Higher prices obtained through the association resulted in a record-breaking harvest in 1924.[19]

Beyond cooperative marketing, the Colorado act enabled the growth of cooperatives for processing and distributing farm products, sharing equipment, managing services, purchasing supplies in bulk, and reselling them to members. But for destroying pests and eradicating weeds, agriculturists relied not on their co-ops but on the county agents to obtain poisons. Campaigns to exterminate rodents and the raptor birds and predatory mammals that fed on them represented perhaps the most notorious example of farmers and ranchers intentionally or otherwise ignoring Liberty Hyde Bailey's admonition to farm in concert with nature and nature's laws. Beginning in 1917, J. C. Hale, the first agent to serve Routt County, spent most of his time helping eradicate the Wyoming ground squirrel (*Urocitellus elegans*). By digging burrows, the squirrel can undermine and destroy cultivated plants and cause range livestock to stumble into burrows, often breaking their legs, which led to their immediate slaughter. Hale reported that within one year, he had mixed, distributed, and overseen the application of 55,000 gallons of Colorado Formula No. 46, a poison created by experiment station entomologist William L. Burnett as the request of the state entomologist Clarence P. Gillette shortly after passage of the Colorado Pest Act in 1915.[20]

Unlike commercially available poisons, Formula No. 46 proved effective when grasses and other forages were still green. Burnett provided county agents with directions for its production: dissolve 1 ounce of strychnine in ½ pint cold water; add ½ pint warm water. Stir in 1 ounce baking soda and ⅛ ounce saccharine; add ½ pint fine salt and ¼ pint petroleum jelly. Put over fire and heat until salt is dissolved, stirring constantly. Remove from fire and stir in ¼ pint flour, making a creamy paste. Pour the solution over 14 quarts of oats, mix thoroughly, and voilà, the poisoned grain is ready for use.[21]

Formula No. 46 was verily a witch's brew that without careful handling could be fatal to humans. Imagine the unease of Floyd D. Moon, a beginning county agent, after he arrived in Routt County in April 1930. Farmers and ranchers had already begun to spread the poison; his immediate task was to prepare more mixture and distribute it where requested. During June he assisted lettuce growers in the vicinity of Yampa, applying the poison to eradicate cutworms as well as squirrels. By August, as nearby hillsides dried up, squirrels moved into irrigated fields and pastures, requiring the application

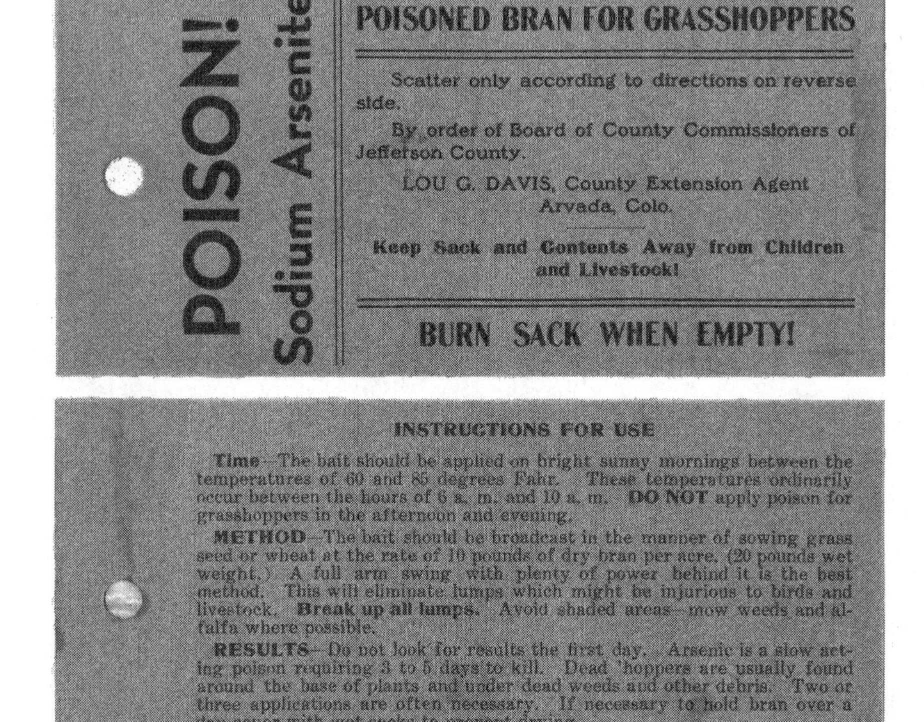

FIGURE 3.5. Field tag, ca. 1930. *Courtesy*, Agricultural and Natural Resources Archive, Colorado State University Libraries, Fort Collins.

of even more poison. During its lifetime, Formula No. 46 was spread over millions of irrigated and non-irrigated acres throughout the state.[22]

Rodents were the principal subject of pest control in Routt County. Black-billed magpies (*Pica pica*) took first place on the list of pests in Boulder County, with extermination of their natural predators contributing to the population explosion. Thickets growing around reservoirs and along irrigation ditches had added to their natural habitat of open spaces, scattered trees, and tangles of bushes near water. Boulder County agent George R. Smith received complaints from farmers and ranchers about magpies killing large numbers of chicken hatchlings, thereby destroying poultry eggs; perhaps most offensive to humans, magpies pecked on livestock sores caused by branding or on live animals caught in barbed wire, forcing ranchers to shelter the affected animals until their wounds healed. Smith publicized the availability of 1-ounce packets of strychnine, instructions included, distributed by the US Biological Survey (today, US Fish and Wildlife Service). Designated pickup sites were banks in Lyons, Niwot, Louisville, Erie, and Lafayette; the chamber of commerce in Boulder; and the extension office in Longmont. During 1927, approximately 250 cooperating farmers and ranchers reduced the county's magpie population by an estimated 10,000, an approach that could hardly be described as adapting to nature.[23]

Unlike strychnine and mixtures such as Formula No. 46, early herbicides were manufactured chemical products. Calcium chlorate, the most popular, was created by dissolving calcium—naturally occurring in certain rocks—in chloric acid under high temperature. New Jersey–based Chipman Chemical Company pioneered its production, direct marketed under the trade name Atlacide, to Colorado farmers starting in the late 1920s. Even with pure seeds and improvements in cultivation, farmers still had to find ways to eradicate noxious weeds, in particular invasive perennials in irrigated fields. Due to the interest generated among farmers by Chipman's sales representatives, the extension service found itself in the position of working hand in glove with a private business to sell a product whose efficacy the experiment station had yet to confirm. In 1929 the experiment station initiated field trials on chemically controlling weeds in fourteen counties; the trials were followed by county agents holding hundreds of spray demonstrations, helping to select equipment, and determining solution strengths and best methods of application.[24]

To further push Atlacide sales, Chipman consigned and delivered the product to county agents who, in turn, sold the product at cost to farmers. Extension agronomist T. G. Stewart defended the agents' intermediary role. Given the apparent effectiveness of calcium chloride in destroying entire weed patches, waiting for experiment station approval would have resulted in doubling and re-doubling the sizes of those patches. Acknowledging that extension could be criticized for joining forces with one particular business, Stewart noted that the service would gladly work with any company that made a comparable product. "We must realize," he wrote, "that noxious perennial weeds are a menace the same as a mad dog and unless something is done about controlling those weed patches some land even though it may be valued at $1,000 per acre will become worthless."[25]

In general, the extension service cautioned against the use of manufactured, that is, synthetic, inorganic fertilizers, holding to the view that the use of pure seed, naturally occurring organic fertilizers—barnyard and green manures—and careful cultivation produced the best results over the long term. To be sure, since ancient times, farmers have used naturally occurring inorganic soil amendments such as lime and gypsum. With better understanding of chemistry and its applications, unprocessed phosphate rock came into limited use during the eighteenth century. Then, in 1841 a learned Irish physician-chemist named James Murray discovered that treating animal bones with sulphuric acid produced a superphosphate that was water soluble for better absorption by crops; he received credit as the first person to market superphosphate commercially. During the course of the nineteenth century, manufacturers began substituting rock phosphate for bone phosphate and developed processes for mixing rock phosphate and sulfuric acid to produce superphosphate in powder form, making that fertilizer easier to transport and spread over fields.[26]

Recall that in 1840 the German chemist Justus Freiherr von Leibig had advanced the proposition that principles of chemistry applied to agricultural practices could dramatically increase yields and reduce costs of production. He theorized that three minerals alone—nitrogen (N), phosphorus (P), and potassium (K)—determined the vitality of crops; those nutrients could be supplied artificially, the exact amount of each depending on analyses of crop ashes and soils in which the crops were to be grown. von Leibig's theory undermined the traditional notion that only decomposed plant materials (organic matter) release nutrients in a way that plants can use. As corollary to his theory, von Leibig adopted what became known as the law of the minimum, a plant's growth depending not on the total amount of essential nutrients available but on the nutrient in shortest supply. Of the three essential minerals, nitrogen was the element most often lacking; thus it was needed in the largest quantities to ensure maximum crop yields. In von Leibig's day, scientists understood that plants naturally derive nitrogen in usable (non-gaseous) form from animal manures, crop residues, and soil organic matter. In addition, legumes fix nitrogen from the air and convert it into a non-gaseous compound evidenced by blood-red nodules on their roots, and precipitation deposits tiny amounts of nitrogen converted into usable form by the atmosphere itself. The implication of Leibig's theory was that synthetic nitrogen fertilizer could artificially extend nature's limits, an exciting prospect consistent with the Enlightenment ideal of the infinite progress of humanity.

With the discovery of nitrogen in the late eighteenth century and the understanding of its value to agriculture, scientists began experimenting with ways to increase its usable supply. It was not until 1909 that the German chemist Fritz Haber perfected the process to convert atmospheric (gaseous) nitrogen into liquid nitrogen. Recognizing the military and industrial applications of liquid nitrogen, the German chemical engineer Carl Bosch led a group of technicians at BASF (Badische Anilin and Soda-Fabrik AG) to successfully transfer Haber's process from laboratory to industrial scale. Known as Haber-Bosch, the process combined atmospheric nitrogen with hydrogen (usually derived from methane, a component of natural gas) into ammonia, which could be spread as fertilizer. It was not until after the end of World War II that American manufacturing plants shifted from making bombs to producing fertilizers—the very time when grain farmers began using seeds of high-yielding hybrid cultivars, which required far more fertilizer than conventional non-hybrid types. Today in farm country, the ever-present cylindrical tanks hitched to the backs of tractors are the containers transporting nitrogen in the form of anhydrous (without water) ammonia, from which farmers spray the fertilizer directly onto their fields.[27]

During the 1920s, farmer purchases of fertilizers, both organic and inorganic, remained negligible. In the San Luis Valley, the extension service did sponsor a few test plots applying ammonium sulfate, an early version of synthetic nitrogen, to head lettuce and sugar beets. County agents reported no measurable improvements in yields. At Fort Collins, meanwhile, bacteriologist Walter G. Sackett had perfected a soil testing method for determining phosphate, potash, and lime content. At least 85 percent of more than 1,000 soil samples taken over a period of three years from farms throughout the state showed phosphate deficiencies. That led the valley Potato Growers Association (est. 1925), which became the statewide Potato Growers Exchange, to recommend use of superphosphate fertilizer, which could be purchased through the exchange. Experiments had shown that the application of superphosphate increased yields and hastened crop maturity; this was especially pertinent for regions that had short growing seasons, where harvesting immature tubers was considered to be a major cause of disease among stored potatoes.[28]

In search of the biggest sales, fertilizer company representatives aimed their most aggressive marketing efforts at sugar beet growers, often misleading them to buy more fertilizer than they needed. Extension horticulturist William M. Case lamented such a waste of money, although he believed growers lost far more money by failing to properly manage their fields. No matter how much commercial fertilizer they spread over their fields, that amendment could not reverse the gradual deterioration of soil fertility caused by loss of organic matter. So concerned were Case and his agronomist colleague T. G. Stewart that they coauthored "A Word of Warning," a leaflet intended for wide distribution by county agents on the use and misuse of all chemical fertilizers. Farmers needed to pay careful attention to the results of soil tests and other field studies before investing in any commercial fertilizers. Even if tests indicated low levels of certain elements, some crops could still produce profitable yields. Commercial fertilizers must not be considered "a cure-all or solution for soil troubles."[29]

Despite technological advances, profit margins kept declining during the 1920s. County agents and farm organizations had helped organize marketing cooperatives for sundry fruit and vegetable growers, grain associations, purebred livestock producers, and dairy operators. Federal and state legislation had greatly facilitated their establishment but did little to secure prices for farm products on par with the prices of products farmers needed to purchase. To wit, a farm wagon in 1913 cost the equivalent of 103 bushels of wheat; by 1923, the cost had risen to 166 bushels.[30] With the problem of overproduction unresolved, the major farm organizations turned to Congress for help. In January 1924, Senator Charles L. McNary (R-OR) and Representative Gilbert N. Haugen (R-IA) introduced legislation that sought to address the problem through increasing demand rather than by reducing supply. Offered as an emergency measure, the McNary-Haugen Farm Relief Bill would have established a temporary US Agricultural Export Commission to purchase surplus production, store it until prices rose, or sell the surplus at a loss on the world market. Under a convoluted system that involved payments in scrip, the commission would recoup any losses by charging producers a fee representing the difference between parity prices and actual market prices.

Even if producers did not receive parity on portions of the crops, their losses would be greatly reduced.

The day after the bill's introduction, the Iowa Farm Bureau Federation endorsed it—which was not surprising since Secretary of Agriculture Henry C. Wallace, co-publisher of the popular journal *Wallace's Farmer* and strong backer of the bill, had helped organize and remained influential in the farm bureau. In 1925 National Farmers Union President Charles S. Barrett brought representatives of twenty-four farm organizations and commodity groups to Des Moines for the purpose of developing a plan to lobby in favor of McNary-Haugen. Approved twice by Congress and vetoed each time by President Calvin Coolidge, McNary-Haugen did serve to publicize the principle of parity, an issue taken up again and again by Congress and favorably acted on more often than not.[31]

The farm lobby's campaign for economic equity, a matter that required political action on the national level, overshadowed a complementary though less tangible and less publicized call on the local level for social justice, espoused by the farmers union and taken up by church leaders on behalf of their rural congregants. To be sure, some like Elmer G. Cutshall, president of Denver's Iliff School of Theology, focused on the need for economic reform, telling the *Colorado Union Farmer* that "until we have redistribution of the means of living of this country so the farmer will be able to live in accordance with the city people, until that time, there will be no hope for the farmer."[32] But more often than not, theologians appealing to rural pastors based their teachings more directly on scripture, in particular on the agriculturist's unique role as steward of the land. It is regrettable, though understandable, not to have found historical records of sermons delivered from the pulpits of rural Colorado churches. But extrapolating back from the present—an occasional necessity with rural history—one can conclude that at least a smattering of rural clergy incorporated stewardship of the land into their sermons, similar to the progressive outlook that had emerged from the country life movement earlier in the century.

The indivisibility of the spiritual and the secular in solving social and economic problems underlay the teachings of the more progressive rural clergy. In 1922 Edwin L Earp, professor of Christian sociology at Drew Theological Seminary, published a series of twenty-four lessons for students preparing to minister in rural communities. In an earlier publication he had recognized the socioeconomic work of the extension service, through county agents and farm bureaus. "But these men," he said, "cannot bring the conviction to the consciences of men like the prophets of God who can add a note of divine authority in saying 'Thus saith the Lord.' 'The earth is the Lord's, and the fullness thereof; the world, and they that dwell therein [Psalm 24:1].'" The role of the rural pastor, Earp continued, is to explain that the land is a gift entrusted by God to the cultivators of the soil for the benefit of all the people: "The land shall not be sold for ever: for the land is mine; for ye are strangers and sojourners with me (Leviticus 25:23)." Furthermore, congregants must be made to feel the guilt of a national sin "if the land is depleted in fertility so that in a few more generations we will be unable to feed adequately our own people."[33] Earp's lessons contained some of the same scriptural passages cited by Liberty Hyde Bailey in *The Holy Earth*. In addition to teaching the moral aspects of land stewardship, Earp believed the rural pastor could extract from scripture certain general insights into cultivation. The ancient Hebrews had established a rest period (the Sabbath) for the land, a primitive version of three-field rotation in his own time: "Now we are learning to conserve the soil values while we intensively cultivate the land."[34]

Adhering to the scriptural injunction on stewardship of the earth did more than ensure good harvests; it ensured that rural communities would thrive. As sentimental and precious as Earp's views may seem today, the fact is that by the early twentieth century, rural churches had been losing influence and membership, with social as well as financial consequences. It is therefore not surprising to read his argument that "the church, the school, and other social institutions can better be maintained when the land produces well and continuously. Depleted soil will not maintain a satisfactory community life, nor will the poor land long maintain a content family life." Again, the rural pastor "must preach and teach the gospel of the sacredness of the soil," leading the congregation "to see the way to soil conservation by teaching better methods of treating soils."[35]

With the election of Herbert C. Hoover as president in 1928, farmers and their advocates had reasons to believe the federal government would take a more active stance on matters agricultural. Born on an Iowa farm and trained as a mining engineer, Hoover had earned a stellar reputation for stepping into the voluntary position of overseeing emergency food distribution in Belgium and other war-torn areas and for serving as wartime director of the US Food Administration under President Woodrow Wilson. Shortly after his inauguration, President Hoover called a special session of Congress to address farm issues. The Agricultural Marketing Act of 1929 embodied a laissez-faire, voluntary approach to achieving economic equity through effective merchandising, contrasted with the proactive, obligatory approach of the McNary-Haugen bill. The marketing act created a revolving loan fund to assist cooperatives in expanding their facilities, to provide operating capital, and to insure them against losses due to price declines; established the Federal Farm Board to purchase, store, and sell agricultural surpluses; and provided for a reporting service on prices, supply and demand, and overproduction.[36]

Historian C. Fred Williams has suggested that the marketing act set forth the ideology underlying all future Republican farm policy and noted that it was developed by William M. Jardine—agronomist, experiment station director, and president of Kansas State Agricultural College before serving as secretary of agriculture under President Coolidge. Jardine believed the farm crisis of the 1920s was a temporary matter, to be righted in time by the markets. While working to improve their cultivation practices, farmers needed to plan their crops according to the prevailing forces of supply and demand. Sometimes, aiming for maximum yields was not the most profitable approach.[37]

Rio Grande County agent Albert A. Goodman could not have agreed more. During a time of declining agricultural prices, he encouraged farmers not to grow too much as an effective way to restore and maintain soil fertility, preserving at least subsistence incomes for their families. He recommended a diversified and balanced cropping system to help maintain soil fertility and control insects, diseases, and weeds. Especially for farmers who grew potatoes, he urged the use of certified or carefully selected seed potato sets. Recognizing that farms in the San Luis Valley did not produce enough high-quality seed potatoes to meet local demand, he urged growers to plant more of them while at the same time planting soil-improving crops such as alfalfa, clover, and field peas. In Goodman's judgment, only those farmers who used quality seed and the most appropriate cultural practices for their particular land would survive.[38]

The October 1929 market crash had little immediate impact on Colorado farmers. Most had not benefited from the Roaring Twenties, which had hidden fundamental weaknesses in the national economy. In 1930 Colorado farmers cultivated a record 6.8 million acres and collected record harvests,

FIGURE 3.6. 'Kanred' wheat harvest, eastern Colorado, 1933. *Courtesy*, Agricultural and Natural Resources Archives, Colorado State University Libraries, Fort Collins.

although some produce was given away because markets did not exist. The state's gross farm income of $183 million represented a decline of roughly 14 percent from the prior year. Falling prices combined with the effects of drought, most severe in south-central and southeastern Colorado, made the farm situation worse. Farm income in 1931 dropped to $126 million.[39]

The year 1932 may have been the cruelest: after a generally mild winter, Colorado experienced late frosts, a series of severe hailstorms, highly destructive cutworm epidemics, and a second year of drought. Wheat prices dropped from 96 cents to 37 cents per bushel, and corn prices declined from 81 cents to 28 cents per bushel; livestock prices experienced similar losses. Farm income fell to $82 million, a 62 percent drop since 1929. Production costs coupled with heavy debt caused more and more farms to be auctioned off. In a desperate effort to keep their neighbors on their land, armed farmers appeared at auctions, threatening violence if auctioneers did not accept their offers, which amounted to pittances. Crowds stormed court houses seeking to destroy land and property tax records.[40]

Meanwhile, the American Farm Bureau Federation and the National Farmers Union had failed in their attempts to obtain relief from the US Congress and the president. Enter Milo Reno, a firebrand populist who grew up on an Iowa farm, became an organizer of farmers, and served as the state's farmers union president during the 1920s. Reno had long argued that the only way farmers could raise prices was by withholding their products from the market. He had hoped to enlist the support of the National Farmers Union but its leadership declined, afraid that a farmers' strike would severely damage its growing network of farm cooperatives. Absent government relief and farmers union support, Reno created the Farm Holiday Association in Des Moines in May 1932—a militant and disorganized movement that garnered thousands of supporters, including some in eastern Colorado. Farmers who did not join the movement were forcibly prevented from moving their products to market. But for many farmers and ranchers it was already too late. Legendary are the cases of entire families abandoning their farmsteads, even causing some small communities to wither away. As the New Deal took hold, the holiday association faded away except for Reno, who in 1937 tried unsuccessfully to convince Jim Patton, then president of the Colorado Farmers Union, and other farm leaders to use the holiday association as the conduit for consolidating farm groups under the aegis of the Congress of Industrial Organizations (CIO).[41]

Dire conditions in drought-stricken counties understandably deflected county agents from their normal work. During 1931, agents organized county relief committees, assisting the American Red Cross in identifying and reaching 20,000 residents of eastern Colorado who required immediate food assistance for themselves and feed for their livestock. Home demonstration agents intensified their efforts to assist farm families, even those who still raised commercial crops, to grow their own home vegetable gardens as part of a program of "making the farm produce a living first."[42] Statewide, as the economy continued to decline and unemployment grew worse, financial contributions to voluntary agencies dwindled. Even the well-established Denver Community Chest, the nation's first United Way chapter, had to reduce its services at a time of maximum need.

As late as 1932, most Colorado political leaders still believed relief was not a government responsibility. A number of self-help groups did emerge spontaneously. Within a period of two months, the Unemployed Citizens' League of Denver attracted more than 30,000 unemployed individuals and their families, including some forced to leave their homes in rural Colorado. The league's principal reason-for-being was to collect and distribute food to members. It organized and sent out corps of volunteers to harvest vegetables and fruits for farmers who, given depressed prices, could not afford to pay for labor. In exchange, farmers gave away some produce, selling the remainder below cost and using the proceeds to purchase necessities for their own families. The extension service helped the unemployed start and cultivate an estimated 15,000 kitchen gardens in forty-two counties. Municipalities, both urban and rural, cooperated by providing space for plots, utilities, equipment, and seeds at no cost to the gardeners.[43]

US senators Edward P. Costigan (D-CO) and Robert M. LaFollette Jr. (R-WI) introduced but failed to get passed a bill to grant financial assistance to the states for their respective relief efforts. During his 1930 senatorial campaign, Costigan had been aided by two men who would take leadership positions in agriculture: Charles F. Brannan, recent University of Denver law graduate and future secretary of agriculture, and James G. Patton, former teacher and future president of the Colorado and national farmers unions. In January 1932, with Democrats in control of the US House of Representatives and a coalition of Democrats and moderate Republicans in the US Senate, President Hoover received and signed the bill creating the Reconstruction Finance Corporation, an effort to shore up public confidence in business by making available loans to banks, farm lending associations, railroads, insurance companies, and mortgage companies. In July, Senator Costigan successfully co-sponsored the Emergency Relief and Construction Act, extending the authority of the Reconstruction Finance Corporation to make loans to the states for direct assistance to farmers, ranchers, and the destitute as well as for public works projects that over a fixed period of time would return to the government the amount of the loans. The act also earmarked loans for construction projects at two federal installations in Colorado: Fitzsimons Army Hospital and Fort Logan.[44]

Progressive in many ways, President Hoover's preference for voluntary action over government mandates may have discouraged him from taking more imaginative and decisive responses to achieve economic and social recovery. In June 1932, his Federal Farm Board announced that its efforts to overcome low farm prices through cooperative marketing had failed. The board recommended that consideration be given to the introduction of federal legislation for the purpose of controlling agricultural production. Five months later President Hoover lost his reelection bid in a landslide to New York governor Franklin D. Roosevelt.[45]

4

Federal Engagement in Agriculture

Through the Emergency Relief Appropriation Act of 1935, Baca County agent Fred C. Case requested funding for a wind erosion remediation project to grant 30 cents per acre for resident land and 60 cents per acre for non-resident land; he was informed that he might receive at most 10 cents per acre. Conditions at the time were bleak. His was among numerous descriptions: "Following the dry winter [1934–35] came a windy spring. The soil was dry as powder and very light, and every little breeze brought a dust storm. Day after day the atmosphere was filled with dust." Fifteen dust storms blew through Baca County during April, the worst occurring on the 15th: "At 3:30 p.m. it was as dark as the darkest night. Dust was so thick it was difficult to breathe . . . People suffered materially. Two Red Cross hospitals were set in operation to take care of the sick. Many people contracted measles and pneumonia."[1]

At one point, roughly one-third of all Baca County residents depended on government relief and recovery funds. Although his principal role was as local relief administrator, Case continued to press for soil conservation measures. Thanks to the Cooperative Farm Forestry Act of 1924, he was able to inform

https://doi.org/10.5876/9781646422050.c004

FIGURE 4.1. Dust storm, Baca County, 1935. *Courtesy*, Agricultural and Natural Resources Archive, Colorado State University Libraries, Fort Collins.

private landowners that bundles of seedlings for windbreaks and shelterbelts were available at his office for as little as 1 cent each, a policy for which he received his share of criticism. One farmer responded that, yes, his land was blowing away, but planting trees was just a "lot of bunk" perpetrated by outsiders who knew little about the land: "Mr. Agent, I been living here since 1912. I think I know more about this land than you do. First you know if you plant trees here they will die. This land won't grow grass leaving out the trees. Why don't you plant trees and show us they will grow. You got lots to learn yet. Good Bye." Case enclosed that unsigned letter with his annual report, no doubt confident that his superiors would recognize the complaint as consistent with a mentality opposed to outsiders and fearful of change.[2]

No county in Colorado has experienced more dramatically the vagaries of weather as well as uses and abuses of the land. Encompassing 2,600 square miles of rolling plains and canyons bordering Kansas on the east and Oklahoma and New Mexico on the south, Baca County averages 17 inches of precipitation annually; that number obscures the fact that, climatically, there is no such thing as a normal year. Recognizing the realities of farming and ranching on the High Plains, the federal government had raised the homestead allocation from 160 acres to 320 aces in 1909 and to 640 acres in 1916. To meet wartime demand for food at home and abroad, the US Department of Agriculture (USDA) had promoted the transformation of grasslands into grain fields. Completion of the Santa Fe Railroad track through the county in 1926 helped facilitate record production of grain and livestock. Reflecting economic growth and decline, the US Census reported a county population of 2,516 in 1910, 10,570 in 1930, and 6,207 in 1940. Under the New Deal, the government imposed soil conservation practices on land recklessly cultivated and bought up parcels that never should have been plowed. Today, 220,000 acres of lands deemed unfit for cultivation make up the Comanche National Grasslands; this constitutes about 13 percent of the county area, managed primarily for grazing by the US Forest Service. The county is now home to fewer than 3,600 inhabitants, roughly one-third the population when Franklin D. Roosevelt became president.[3]

In presenting a "New Deal for the American people," President Roosevelt rejected the measured, mostly voluntary approach to economic relief that had failed under the past administration. Instead, he called for swift federal action to revitalize the economy. During his first 100 days, the US Congress passed legislation for emergency relief, temporary recovery, and systemic reform to protect against another economic disaster. Because of its practical, non-ideological approach, the New Deal remained a work in progress. The administration willingly risked experimenting with programs, abandoning those that failed and adopting new ones in their stead. Whether for relief, recovery, or reform, all programs affected agriculture to some extent. Generally speaking, farmers and ranchers reluctantly accepted the necessity of federal engagement, reminding us of Bernard DeVoto's famous quip that to this day describes the prevailing western attitude toward the federal government: "get out, and give us more money."[4]

In an effort to stanch the financial panic, President Roosevelt suspended all banking transactions within thirty-six hours of entering office. Three days later, on March 9, 1933, Congress passed the Emergency Banking Relief Act, which authorized the Department of the Treasury to close insolvent banks and the Federal Reserve to supply currency to banks reopening under treasury's supervision. Except for eight small banks located in rural communities, banks in Colorado resumed business, with public confidence restored. Next, Congress approved the creation of federally financed job training and

FIGURE 4.2. CCCers constructing contour ridges near Springfield, 1935–40. *Courtesy*, Colorado State Archives, Denver, #60041, ff. 5, col. 181.

temporary jobs for the unemployed. The best-known such program was the Civilian Conservation Corps (CCC), which employed single men ages eighteen–twenty-five to help conserve the nation's forests, fields, and parks. During its ten-year existence, the CCC in Colorado created temporary jobs for 30,000 unemployed young men recruited primarily from within the state, instructed them in useful job skills and in subjects that allowed them to complete their secondary education, and provided them with healthcare. While the army managed the camps, the Departments of Agriculture and the Interior planned and selected the work projects. At first, the CCC worked primarily in Colorado's national forests, benefiting graziers by building roads, bridges, and stock-watering systems; preventing and fighting fires; eradicating predators; controlling insects and disease, and much more. Beyond emergency relief, the Taylor Grazing Act (1934) and the Soil Conservation Act (1935) enabled the CCC to work on more permanent programs to restore and maintain range and farmlands.[5]

The Federal Emergency Relief Act (FERA) of May 21, 1933, authorized grants to the states for redistribution to public agencies and charities either for the purpose of creating jobs or making direct payments ("welfare") to the needy. In Colorado the relief committee established during the Hoover administration had become known as the Colorado Emergency Relief Administration. The committee entrusted the extension service with setting up county and community committees to encourage the planting of subsistence gardens, with county agents helping farm and ranch families become as food self-supporting as possible. Public perception to the contrary, the Roosevelt administration always favored work programs over welfare and, whenever possible, partnership funding over federal grants alone. FERA did require states and local communities to put up 3 dollars for every federal dollar received; but the Colorado legislature, even with a Democratic majority, declined to act on the matching requirement. After the federal relief administrator, Harry L. Hopkins, interceded on Colorado's behalf, FERA monies did flow into the state, bearing more than 80 percent of relief costs by late 1933. In addition, FERA distributed surplus foods purchased by the Federal Surplus Relief Corporation, the New Deal's version of President Hoover's Federal Farm Board. Federal distributions far exceeded donations of surplus crops made by Colorado farmers.[6]

The New Deal initiatives with the greatest impact on agriculture derived from passage of the Agricultural Adjustment Act (AAA) on May 12, 1933, and its successor, the Soil Conservation and Domestic Allotment Act, on February 29, 1936. The AAA confirmed official policy on parity: "to establish and maintain such balance between the production and consumption of agricultural commodities, and such marketing conditions therefor, as will reestablish prices to farmers at a level that will give agricultural commodities a purchasing power with respect to articles that farmers buy, equivalent to the purchasing power of agricultural commodities in the base period [August 1909–July 1914]."[7]

To achieve parity quickly, the federal government paid farmers to reduce production, in Colorado of wheat, corn, hogs, and milk; initiated voluntary agreements with processors, farm cooperatives, and other handlers of agricultural products to regulate marketing; and levied a special tax on processors of basic commodities to cover the costs of paying farmers not to plant, to expand markets, and to eliminate surpluses. At the same time, the AAA sought to protect consumers from excessive price increases by keeping retail expenditures for agricultural products on a relative par with what farmers received for their products during the base period.[8]

Once again, the federal government looked to the extension service to oversee implementation of federal agricultural policy. County agents were charged with finding as many farmers and ranchers as possible to participate in acreage reduction, making sure that subscribers complied with rules and regulations. With understandable reticence, county agents became less teachers and advisers and more mentors and supervisors.[9]

Wheat was the first commodity crop in Colorado to come under the acreage reduction program. Initially, growers opposed AAA allotments because they were based on acreages already reduced by drought. After the administration adjusted allotments upward, about 90 percent of growers signed their AAA contract; wheat acreage declined by 15 percent at the time when 'Tenmarq' was replacing 'Kanred' as the red winter wheat cultivar of choice. In fall 1933 the government extended the AAA program to corn and hogs, although sorghum had begun to replace corn as the preferred forage crop. At the Akron experiment station, agronomists Joseph F. Brandon and David W. Robertson had developed two new sorghum cultivars, 'Improved Coes' and 'Highland Kafir.' In a controversial move to reduce production, the federal government bought up piglets, then had them slaughtered, turned into salt pork, and distributed to the needy by the Federal Emergency Relief Administration. Most producers participated in that program, reducing corn acreage by 20 percent and hog production by 25 percent. During the drought years 1933 and 1934, when most seeds sown failed to germinate, AAA payments represented the only source of cash income for many farmers and ranchers, in essence a form of emergency relief.[10]

At the same time the federal government sought to increase farm income through production allotments, it addressed the continuing scarcity of affordable farm loans. The Farm Credit Act of March 27, 1933, authorized

FIGURE 4.3. Corn in drought-affected field, 1941–46. *Courtesy*, Agricultural and Natural Resources Archives, Colorado State University Libraries, Fort Collins.

creation of the Farm Credit Administration—a consolidation of existing federal programs, including President Hoover's Federal Farm Board. The new agency could make emergency loans for supplies and equipment, as well as purchase outstanding farm mortgages and refinance them at lower interest rates and for longer periods of time. In the case of corn grown on reduced acreages under AAA contracts, the Farm Credit Administration made available non-recourse loans: if income from a harvest was insufficient to repay a loan in full, the government agreed to absorb the loss and seek no further compensation from the farmers.[11]

Sugar is not generally perceived to be a basic commodity, but by the late 1920s sugar beets had become Colorado's largest cash crop, a major contributor to the state's economy. As prices declined 25 percent between 1929 and 1933, farmers sought to compensate by increasing production by 50 percent; cash-strapped consumers, however, bought less sugar. Pressed by sugar cane and sugar beet refiners, respectively, Representative John M. Jones (D-TX) and Senator Edward P. Costigan (D-CO) successfully introduced an amendment to place sugar production under AAA control. In addition, the

FIGURE 4.4. Thinning sugar beets, 1939. *Courtesy*, Agricultural and Natural Resources Archive, Colorado State University Libraries, Fort Collins.

Jones-Costigan Act of May 9, 1934, addressed the much publicized mistreatment of sugar beet workers and their families. It required beet farmers to pay their laborers more and reduce work hours, and the act prohibited the employment of children under age fourteen.[12]

The sad truth about the sugar beet industry was its dependence on unskilled, low-paid, mostly Hispanic seasonal farm labor. Growers were further squeezed by lack of competition among refiners. Since transporting beets—which are bulky and heavy—over long distances made no financial sense, growers had no choice but to sell to the closest factory. Even with grower-processor contracts, farmers were continually pressed for more efficiency, which translated into paying laborers as little as possible under the law. One way growers tried to maintain profits was by contracting with heads of migrant families for cultivating and harvesting specific acreages rather than by outright employment. To fulfill their contractual obligations, heads of families needed to have their dependents work in the fields. During spring 1933, nearly half of all relief monies distributed in Colorado went to the beet-producing counties. By supplementing substandard earnings with welfare payments, the federal government effectively subsidized the growers.

It did so in another way, too. Despite their opposition to acreage control, beet growers overwhelmingly welcomed government payments for fields not planted. By the end of 1935, beet growers were collecting 40 percent of all federal in-lieu payments going to the state, helping to offset losses in tax revenues experienced by local governments.[13]

From the perspective of New Deal economic and social reformers, there was nothing inconsistent about the government paying some farmers to reduce production and resettling other farmers to improve productivity. Sharing the idealism of some of the early union colonists, New Dealers entertained the idea of resettling destitute farmers into planned communities where they could grow most of their food, earn extra income by taking jobs in nearby towns, exchange services among themselves, and participate in civic life. Starting in 1935, the Resettlement Administration purchased several hundred thousand acres of submarginal land in northeast and southeast Colorado, which generated fear that the federal government meant to return most cultivated dry land back to grazing. In 1936 the government did move several hundred families to three new government-sponsored towns on the Western Slope. Like all utopian experiments, these failed, too.[14]

While complaining about the intrusion of the federal government into every aspect of agriculture, farmers did take advantage of crop reduction payments and the benefits provided by the many "alphabet agencies," so-called because of the acronyms used to identify them. Because of the rapidity with which new programs started, the stumbles encountered en route, and apprehension about change, certain AAA-sponsored projects generated discontent and even outright opposition that went beyond small farm communities. On January 6, 1936, the US Supreme Court in *United States v. Butler* (receivers of Hoosac [cotton] Mills) declared that federal control of production under the AAA and the tax assessed on commodity processors to pay farmers for not planting were unconstitutional. The Roosevelt administration quickly came up with an alternative approach—tying federal aid to soil conservation—designed to accomplish the same purpose of reducing production and raising prices; this approach had unanticipated and vastly significant results for conservation of land and water over the long term.[15]

With the demise of the AAA, certain reforms to strengthen the rural economy fell by the wayside, but President Roosevelt excelled in broadly interpreting emergency relief legislation to achieve far-reaching socioeconomic results. By authority of the Emergency Relief Appropriation Act of 1935, the president issued the executive order that created the Rural Electrification Administration (REA). The Rural Electrification Act of May 20, 1936, clarified and augmented the role of the REA, providing for loans and other forms of assistance to consumer-owned cooperatives to furnish electricity to rural areas. Although county agents in Colorado had been assisting individual farm and ranch families in electrifying their operations since the mid-1920s, the REA provided for entire regions to receive power from central transmission stations. The first REA cooperative in Colorado, Grand Valley Rural Power Lines, began operations in 1937 around the towns of Grand Junction and Fruita. In 1934 about 11 percent of Colorado farms and ranches were getting electricity from a central source; by 1943 that number had risen to 44.5 percent, served by nineteen separate co-op associations. The impact of electrification on household and farm chores was immense, but it did more than that. In the opinion of Routt County agent Edson W. Barr, it gave rural residents an equal standing with urban residents and offered encouragement to rural leaders seeking to retain young people on the land. Today, twenty-two electric co-ops distribute electricity to rural, ex-urban, and suburban Colorado.[16]

The effects of drought on cultivated lands in eastern and southeastern Colorado may have attracted the most attention, but drought had also caused severe damage to rangelands, especially in northwestern Colorado. Of the 67 million acres that make up the state, 40 million are designated as rangelands. Subsequent to the Forest Reserve Act of 1891, 10 million acres of un-appropriated, primarily mountainous land were withdrawn from sale and placed under federal management. By the early 1930s, after accounting for sale of rangelands to private parties, 7 million acres remained unreserved and unmanaged. To increase the carrying capacity of rangelands, regardless of ownership, scientists connected with the experiment station promoted the replacement of native sagebrush with cultivated grasses. Preferred species deemed more digestible and nutritious for livestock were smooth brome (*Bromus inermis* Leyss.), slender wheatgrass (*Elytrigia elongata* [Host.] Nevski), and crested wheat (*Agropyron cristatum* [L.] Goertn.). Extension agents recommended burning as the least expensive way to eliminate sagebrush before cultivation and seeding of the grasses. The opening of the Moffat Tunnel scheduled for February 1928 had raised expectations for transporting livestock to markets by rail as well as fears of overgrazing in northwest Colorado. In anticipation of land-use conflicts, the extension service had sponsored a conference on agricultural economics in Steamboat Springs in October 1927. Range and forage section participants expressed opposition to increasing the numbers of range livestock pending more precise estimates of carrying capacities; instead, they favored range improvements such as re-vegetation of lands already overgrazed and advocated for federal management of public rangelands.[17]

Seven years later, severe deterioration of rangelands caused by prolonged drought compelled US representative Edward T. Taylor (D–Glenwood Springs) to introduce legislation mandating federal management of public rangelands. Because of his immense popularity among stock growers and wool growers, Taylor had managed to overcome their resistance to federal control of the land. The Taylor Grazing Act of June 28, 1934, is celebrated as one of the nation's most significant pieces of conservation legislation. Its overall purpose: to promote the highest use of public lands, consistent with their preservation. The act sought to address issues of immediate concern to Taylor's Western Slope constituents, foremost of which were overgrazing and soil deterioration that can be seen still today at higher elevations. Seasonal movement of large flocks of sheep from Utah into the mountains of Colorado contributed to making plant regeneration difficult, if not impossible; unlike cattle, sheep graze in tight bands and forage close to the ground, which explains the need to move them frequently. As had occurred earlier on the eastern plains, purchases of strategically placed homesteads by large operators using legal loopholes had closed off certain rangelands to other users.[18]

In addition to ending the federal policy of selling public lands to private parties, the Taylor Grazing Act authorized establishment of the US Grazing Service (since 1946, the Bureau of Land Management) to manage unappropriated public rangelands. Assistant Secretary of the Interior Oscar L. Chapman, a protégé of Senator Costigan, appointed Routt County rancher Farrington R. Carpenter, a Republican and also a Costigan protégé, to be the first director of the US Grazing Service. Although Carpenter occupied that position only briefly, he did begin the process of establishing local grazing districts and recruited local ranchers to serve on district committees to provide advice on fees, access, and other management issues. Under supervision of the US Grazing Service, the CCC operated between five and eight summer work camps annually on the Western Slope. Corpsmen engaged in building stock ponds and corrals; destroying noxious weeds, pests, and larger wildlife considered objectionable by ranchers; reseeding overgrazed lands; rip-rapping stream banks; and building stock ponds and other improvements for the care and management of permitted livestock.[19]

At first, New Dealers viewed soil conservation principally as a form of job creation within the grand scheme of relief, recovery, and reform. The National Industrial Recovery Act of June 16, 1933, had provided emergency funds to conserve and develop natural resources. In carrying out the act, interior secretary Harold L. Ickes established the Soil Erosion Service. He appointed as director Hugh H. Bennett, soil scientist in the USDA Bureau of Chemistry and Soils and longtime crusader for soil conservation. Bennett and US Forest Service grazing inspector William R. Chapline had published an influential essay, "Soil Erosion: A National Menace," which prompted Congress to appropriate funds for soil erosion surveys and for ten regional soil erosion experiment stations beginning in 1929. The nearest station to Colorado was located on the grounds of the Kansas agricultural experiment substation at Hays.[20]

Dust clouds originating on the High Plains darkened the skies over Washington, DC, and the northeastern states during 1934 and early 1935. That provided director Bennett with the backdrop for convincing Congress to convert the Soil Erosion Service into a permanent conservation agency. Using some of the language in the Bennett-Chapline essay, the Soil Erosion Act of April 27, 1935, placed all federally supported erosion control activities under the auspices of the Soil Conservation Service (SCS), a new permanent division within the USDA. Secretary Henry A. Wallace (son of Henry C.) appointed Bennett as the first chief, a position he held until 1951. In contrast to prior legislation that focused on erosion control, the Soil Erosion Act enabled the SCS to carry out erosion prevention activities, including work on private properties if voluntarily given permission to do so; to cooperate with and make grants to other agencies, governmental or otherwise; and only in rare cases to condemn land in an effort to conserve soil. Just as the extension service was based on a cooperative relationship between the agricultural college and rural residents, the genius of the SCS entailed a cooperative relationship between the USDA and rural residents. In complementary ways—through education, technical assistance, and financial assistance—the publicly funded institutions shared risk with their agricultural constituents.[21]

Following passage of the Soil Erosion Act, the SCS conducted aerial mapping surveys with the aim of determining best farming practices for areas of similar topographic and climatic conditions, inaugurated widespread reseeding of the most eroded lands, and established demonstration projects in collaboration with the extension service and the CCC. In southeastern Colorado, the contour plowing that Joshua Adams had pioneered at Cheyenne Wells became the farming technique of choice. Considering that agriculturists had long excelled in making custom equipment out of scraps, A. E. McClymonds in Baca County converted a plow to make furrows on the contour. Charles T. Peacock in Elbert County invented a machine to create tiny temporary check dams every few feet across contour furrows to capture moisture. In late 1935, a group of Pueblo County farmers calling themselves the Last Man's Club petitioned the federal government to subsidize contour plowing. Secretary Wallace endorsed their proposal, which was incorporated into the Soil Conservation and Domestic Allotment Act of February 29, 1936.[22]

Born out of the national emergency, tempered by the Supreme Court decision on the Agricultural Adjustment Act, and written by New Dealers committed to economic and social reform, the Soil Conservation and Domestic Allotment Act set forth, in the words of President Franklin D. Roosevelt, "a long-time program for American agriculture."[23] The new act put forward three main objectives: to restore, maintain, and improve soil fertility; to establish and continue parity of farm income; and to guarantee consumers the ready availability of agricultural commodities at fair prices over the long term. Instead of offering farmers incentives to take land out of production, the act incentivized farmers to convert their fields from soil-depleting crops in surplus, such as corn and wheat, to soil-building crops that were in demand, such as legumes and forage grasses. Unlike the AAA, the Conservation and Domestic Allotment Act prescribed neither contracts nor production quotas; it did offer payments to farmers on evidence of conservation practices, as verified by county agents. The expectation was that conditional payments would carry farmers through poor crop years; payments were to come out of general appropriations, not from a special tax. The act made official the policy of income parity: "the ratio between the purchasing power of the net income per person on farms and that of the income per person not on farms that prevailed during the five-year period August 1909–July 1914."[24]

Despite the New Deal's pragmatic outlook, one cannot help but suggest that the idealists of that administration viewed conservation not only as practical but also as the right approach ethically. Among those idealists was a young USDA lawyer named Philip M. Glick. To convince farmers and ranchers, ever suspicious of federal intrusion, to conserve their own soils, he devised a novel method: engage those farmers and ranchers at the most local level in the planning, direction, and execution of conservation activities. Since the granting of federal funds was conditional on acceptance by the individual states, Glick helped draft a standardized text of a state soil conservation district law, which President Roosevelt sent to the nation's governors for consideration by their respective legislatures.[25] In 1937 the Colorado General Assembly (state legislature) passed and Governor Teller Ammons (D) signed the Colorado Soil Conservation District Act. The act acknowledged that the state had lost approximately 6 million acres (one-tenth of its total area) to erosion, that these losses had been caused by improper farm and range practices, and that such losses would increase "until and unless a uniform method of land use providing for the conservation and preservation of natural resources . . . is established by law over the entire state (CRS 35-70-101)."

The act authorized the establishment of a statewide Soil Conservation Board (now part of the Colorado Department of Agriculture) to oversee the creation and maintenance of soil conservation districts. To establish a district, at least twenty-five resident farmers and ranchers had to petition the board, which convened public hearings to set boundaries, followed by a voter referendum. If a majority of residents in a proposed district voted to approve, then the state board appointed two supervisors and residents elected three supervisors—all district residents—to oversee the district. Once established as a publicly supported corporate body, district supervisors were invested with broad authority: to accept contributions of money, services, and materials; hire personnel to provide technical assistance; and contract with farmers and ranchers who agreed to practice soil conservation in return for such assistance as well as for loans to purchase equipment and supplies. The district board could enforce its own regulations, seeking relief through the courts if a resident failed, for example, to control wind erosion that caused damage to a neighbor's property. Colorado's first soil conservation district was the Great Divide Soil Conservation District (Moffat County), established in 1937, followed by the Western Baca and Smoky Hill River (Kit Carson County) Districts.[26]

The Colorado act required that membership in the statewide Soil Conservation Board include the directors of the extension service and the agricultural experiment station, the state commissioner of agriculture, and an appointee of the secretary of agriculture. Convinced that extension retained close ties with the farm bureau, which was unsympathetic to the New Deal, Secretary Wallace overruled his advisers; he insisted that soil conservation district supervisors, not extension agents, direct district activities. With the advantage of hindsight, historian Douglas Sheflin has cautioned that insufficient credit was given to extension's long-standing efforts to build trust between the federal government and agriculturists. Following the president's declaration of economic emergency, Colorado extension director Fridtjof A. Anderson had organized a committee of representatives from every federal and state agency that dealt with matters agricultural to meet monthly to coordinate their activities. Once again, thrusting county agents into administering federal relief and recovery had placed them in awkward positions.[27]

Whether in the Dust Bowl area or elsewhere in the state, most Colorado farmers took advantage of available government technical and financial assistance. Based on his review of several hundred applications for loans and outright grants between 1933 and 1937, Rio Grande County agent Albert Goodman concluded that the majority of requests resulted not so much from drought as from decades of flawed farm practices. Time and again, county agents revealed their frustrations with farmers who, for whatever reasons, declined to practice cultivation that kept soils fertile and made harvests more profitable. Most farmers preferred to gamble year after year with the same cash crops. Among 235 farms registered for the soil conservation program in the vicinity of Sargents (Saguache County), only 35 percent grew alfalfa or clover and 42 percent planted 10 acres or less in soil-improving crops. As a missionary for best practices, Goodman lamented that among agricultural lenders, only the Federal Land Bank of Wichita required loan recipients to participate in a soil conservation program. In neighboring Conejos County, extension had held numerous educational meetings to explain the benefits and requirements of conservation programs. Agent C. M. Knight lamented that by late 1939, fully half of all farmers had still failed to sign up for any federal program, thereby forfeiting valuable financial and technical assistance such as preparation of written documentation of conservation plans customized to each farmer's operation.[28]

During the 1934 and 1935 seasons, farms in eastern Colorado harvested virtually nothing. With the coming of spring 1936, through the good offices of county agents, the Soil Conservation Service began an emergency program to pay 20 cents per acre to farmers who contour-plowed their fields, lent those farmers cash so they could operate their tractors, and secured pledges from local grain elevator operators to extend loans only to farmers who contour-listed. Cheyenne County agent C. L. Hart heard from farmers that the contour-plowing program was more valuable to them than any federal program yet administered: "It halted the blow situation, [it] protected adjoining land that was not listed, machinery was put into shape for the planting season, [it] helped many to get some crop planted, [and it] started the movement toward contour farming and put hundreds of blow fields [barren of vegetation] in a condition to retain enough moisture to grow a good coverage of weeds to hold the land in the future." During that first year, 92,000 acres—about 20 percent of cultivated land in Cheyenne County—had been part of the program, which also protected an estimated additional 150,000 acres of adjoining land.[29] By then, county agents and SCS staff everywhere could base their recommendations on timely research provided by those scientists who had applied their research to addressing the requirements of specific sites. In 1936 two agronomists, Joseph Brandon (USDA) at the Dry Land Experiment Station near Akron and Alvin Kezer, published perhaps the best and certainly the most readable study to that date on the causes of soil blowing and remedies for its control. Federal incentives had made it possible, indeed profitable, for growers to be "ruled by the soil-blowing protection needs, not by the current market value of crops."[30]

To carry out its initial projects, the Soil Conservation Service depended on the Civilian Conservation Corps. Beginning in 1935, the CCC set up a series of camps in eastern and southeastern Colorado on lands leased from farmers. From the camps, corpsmen fanned out within 25-mile radii to work on demonstration plots located on private lands. Farmers provided what they could in the way of supplies and materials, and corpsmen used government equipment to do the actual work: contour plowing, listing, and strip cropping; building contour ditches and check dams; rip-rapping stream beds; reseeding overgrazed areas; planting windbreaks; and engaging in an array of other conservation activities. Since many of the corpsmen came from farm families, the expectations were that when they returned to their own farms, they would put into practice what they had learned on the demonstration farms.[31]

To outsiders passing through, the demonstrations farms appeared as islands of recovery and restoration in a vast wasteland. In preparing a guide to the American West, author Dorothy C. Hogner took a detour through Baca County to describe conditions to her lay readers in the East and provide some flavor of how local people were coping with the aftermaths of drought. Approaching Springfield, the county seat, she noted barren countryside—no trees, no bushes, no greenery—dotted with abandoned homesteads. The only living things were grasshoppers and birds, the latter escaping the summer heat "with wings spread, crouching, panting" in the thin dark shadows of telephone poles. Stopping at the SCS office, she learned from the resident soil scientist that the agency had leased 40,000 acres from landowners for five years for demonstration projects. His technical staff consisted of an agronomist, an engineer, a range management expert, and a "contact man" acting as intermediary between the SCS and farm and ranch lessors. Driving Hogner to a field day at the demonstration site, the soil scientist submitted that the dust storms had been caused by overgrazing, cultivating land that should never have been plowed, and the fact that 45 percent of all private land belonged to absentee landlords—speculators who failed to care for their land or their neighbors during years of extreme drought. At the field day site, Hogner joined forty farmers and ranchers from a neighboring county present to inspect green fields of sorghum and corn contour-listed and cultivated by CCCers under SCS supervision. The visitors also inspected recently planted trees in roadside ditches and saw farmhouses—once buried under dunes—excavated, refurbished, and reoccupied.[32]

Even as conditions improved, Baca County agent Raymond H. Skitt noted that six years of drought and dust storms, reliance on government relief, and continuing shortages of life's necessities had taken a toll on the human spirit. Among those who had decided to remain or could not move, an air of discontent and apathy prevailed; people were satisfied with just getting along

FIGURE 4.5. Abandoned farmstead, Baca County, 1940. *Courtesy*, Agricultural and Natural Resources Archive, Colorado State University Libraries, Fort Collins.

rather than working to improve their conditions. Winter snows and spring rains combined with greater accessibility to farm credit and more payments for soil conservation helped infuse "new life and hopes in the people." As the economic emergency receded, the federal government turned its resources from relief to recovery. In 1937 the Farm Security Administration (FSA) succeeded the Resettlement Administration, making loans to rural residents who could not get credit elsewhere and thereby enabling them to purchase land, equipment, seeds, and livestock. Families participating in FSA loan programs were eligible to obtain training, education, and even healthcare—all consistent with New Deal policy to improve rural life in general. After an absence of eight years, the Baca County Fair resumed in 1938, attracting half of the county's total population, according to Skitt.[33]

Notwithstanding generous technical and financial assistance from the SCS and even after witnessing the benefits of soil conservation, Baca county residents rejected a referendum to create a county-wide soil conservation district by a margin of three to one. Misunderstanding the governance structure of conservation districts or simply opposed to taxes, naysayers argued falsely that district residents would be taxed without the authority to determine tax rates or how monies were to be spent. Shortly after the referendum failed, voters in two areas of the county did approve the creation of two separate smaller districts. In 1938 directors of the West Baca District issued a ban on plowing up virgin grassland. By valuing soil conservation over maximum production, district supervisors effectively placed the common good for the long term above private interests in the short term. Some property owners interpreted the district's ban as an attempt by government to usurp control of private lands. The courts denied their appeal and sided with the conservation district.[34]

Drought conditions during the 1930s had alerted the general public to the perils of irresponsible land use and caused the federal government to react with decisive remedial action. Drought exacerbated the natural scarcity of water, putting pressure on state government to take the utmost care in supervising the appropriation, distribution, and diversion of state waters. Recall that Colorado was the first state to adopt the doctrine of prior appropriation and the administrative structure for its implementation and that an important corollary of beneficial use was that water must not be wasted. To reduce waste through precise measurement, Ralph L. Parshall, head of the irrigation section at the Colorado Agricultural Experiment Station, had perfected an artificial water chute to measure irrigation flow. Without such accuracy, Parshall wrote, "the appropriator of water cannot make a definite statement as to how much water he actually uses, and if a dispute should arise it would be difficult for him to furnish satisfactory proof of his established rights." To this day, the Parshall flume remains the standard measurement instrument used by irrigators worldwide.[35]

In the minds of water managers, just as critical as water waste was the matter of getting water to where it could be used most beneficially; in the case of Colorado, that meant moving water from the Western Slope to augment existing supplies in the eastern counties. Almost from the beginning of irrigated farming in the South Platte Valley, agricultural promoters with supporting data from the agricultural college had advocated for trans-basin diversion, the major obstacle being the high cost of such projects. The National Industrial Recovery Act of June 1933 earmarked $2 billion to hire the unemployed to work on water conservation, reclamation, and flood control. That stirred northern Colorado promoters into action, foremost among them *Greeley Tribune* editor Charles Hansen. As chairman of the Greeley Chamber of Commerce Irrigation Committee, he brought together

FIGURE 4.6. Ralph L. Parshall measuring water, 1946. *Courtesy*, Agricultural and Natural Resources Archive, Colorado State University Libraries, Fort Collins.

irrigators from Weld, Larimer, Morgan, Logan, Sedgwick, and, later, Boulder and Washington Counties; formed the Northern Colorado Water Users Association (NCWUA); and became the group's first president in early 1935.[36]

Historian Daniel Tyler has meticulously documented the campaign that led to the construction of the Colorado–Big Thompson Project (C-BT), diverting water from Grand Lake and the Colorado River on the Western Slope to the Big Thompson River and northeastern Colorado—with long-term benefits and drawbacks as yet unimaginable for agriculturists and non-agriculturists on both sides of the Continental Divide. Since only the federal government had the resources to undertake such a massive project, the NCWUA and its supporters turned to the relevant federal agency, the US Bureau of Reclamation, directed until 1936 by Elwood Mead whose rise to prominence as a hydraulic engineer had begun in northeast Colorado.[37]

Anticipating that the project would be approved and funded, Western Slope irrigators organized to protect their water rights. In a remarkably orderly series of negotiations over four years, east and west agreed that as part of the C-BT project, new storage reservoirs had to be built on the Western Slope to hold replacement waters sufficient to meet the needs of Western Slope users. Congressman Edward Taylor played a critical role in convincing his Western Slope constituents to approve the project. To ensure federal financial approval of the project, Charles Hansen and the NCWUA successfully petitioned the state legislature to enact statewide water planning. The Water Conservancy Act of May 31, 1937, established the Colorado Water Conservation Board for the purpose of controlling, protecting, and developing all waters in the state. To the issue at hand, the act authorized the establishment of water conservation districts and enabled districts to accept federal grants, incur debt, contract with the reclamation bureau for construction, and collect rent from users taking advantage of a water project's distribution network. In 1938 the NCWUA transformed itself into the Northern Colorado Water Conservancy District (NCWCD), arguably still the most effective advocate for trans-basin diversion.[38]

From its conception, the Colorado–Big Thompson Project generated opposition from a variety of interest groups inside and beyond Colorado. The National Park Service and conservationists feared that drilling a tunnel through the Continental Divide in Rocky Mountain National Park would diminish the nature of the park and set a dangerous precedent for development throughout the national parks system, although the 1915 legislation creating Rocky Mountain National Park had specifically allowed the Bureau of Reclamation such access. Conservationists further argued that massive trans-basin diversion would undermine the Roosevelt administration's concerted efforts to control agricultural production and preserve natural resources. Because sugar beets, the leading commodity crop, required large landholdings to be profitable and with only a few refiners controlling production, the project appeared to treat with disfavor the small-scale farmers earlier progressives and New Dealers had long envisioned as the principal beneficiaries of irrigation in the arid West.[39]

The Colorado–Big Thompson Project posed an especially uncomfortable dilemma for farm organizations with members on both sides of the Continental Divide. As a result, they settled on protecting agricultural waters against other uses but stayed out of conflicts among irrigators. Perhaps the single most vocal opponent of trans-basin diversion was Archer C. Sudan, MD, who had built his own clinic and practice in Kremmling (Grand County).

A South Dakotan with degrees from the University of Chicago and Rush Medical College, Sudan had been attracted to Grand County by its blue-ribbon fly fishing. He argued to the bitter end that by reducing and artificially controlling stream flow, trans-basin diversion would cause irreparable damage to fishing on the Colorado River and its tributaries. His fly-fishing successors argued later that trans-basin diversion irrevocably changed the natural ecosystem. After much debate, Congress authorized funds for the project, with construction to begin in late 1937. The Works Progress Administration began and the Bureau of Reclamation completed construction; the entire distribution system became fully operational twenty years later.[40]

Although trans-basin diversion did not affect water quality, some streams flowing out of the Front Range foothills had long served as dumping places for mine tailings, of special concern to the fledgling nursery business serving the Denver metropolitan area. A case in point, Clear Creek contained enough mine residue to stunt or even destroy certain plants as irrigation waters entered fields. Jefferson County agent Louis G. Davis described the residue as consisting of fine particles that "form a seal coating on top of the soil and prevent water from soaking into the soil as well as placing hundreds of tons of material on the good soil that is detrimental to plant growth." The extension service collected evidence that would be used by pioneer nurseryman William W. Wilmore and four fellow growers from Wheat Ridge (Jefferson County), convincing the legislature to pass protective stream legislation in 1934. After losing on appeal to the state supreme court, the mine companies lobbied legislators, again without success, to enact legislation to overturn the court's decision—all the while continuing to dump tailings into Clear Creek. In 1935, Wilmore won a claim for damages that resulted in one company being fined today's equivalent of $10,000. It is unclear if that company paid and ended its dumping.[41]

Water and soil conservation, meanwhile, retained its priority with the Roosevelt administration. The reauthorized Agricultural Adjustment Act of February 16, 1938, was the last major piece of New Deal agricultural legislation and its most comprehensive statement of farm policy. To qualify for federal payments or federally backed loans, farmers had to show how they planned to practice conservation. The act authorized special payments to those who voluntarily agreed to retire and restore land found unsuitable for cultivation; farmers who withdrew land but did not restore it were

FIGURE 4.7. Wheat field, Akron, 1936. *Courtesy*, Agricultural and Natural Resources Archive, Colorado State University Libraries, Fort Collins.

fined and further payments were withheld until they complied. As with prior federal legislation, the overall aim was to raise farm prices and farmer income, but this time through control of marketing rather than control of production. Farmers participating in federal loan programs could withhold part of their crops in abundant years and release them in lean years, maintaining reserves to ensure consumers of adequate supplies of basic commodities. By supplementing direct payments with a subsidy meant to bring farmer income closer to parity and by inaugurating federal crop insurance, the government moved closer to guaranteeing economic security for farmers. Initially, crop insurance covered only wheat, with premiums payable in bushels of wheat. By the fall of 1941, 6,000 Colorado wheat growers were participating in the program.[42]

To bring the state into conformance with federal legislation, the general assembly passed the Colorado Agricultural Marketing Agreement Act of 1939. Since the founding of the Grand Junction Fruit Growers Association in 1891, producers of perishable fruits had tried to band together to control prices at harvest time when the bulk of their produce came on the market. Because such associations were voluntary, some producers chose not to join, believing they could do better on their own; this enabled chain grocers to buy surpluses at depressed prices, mark them up, and still sell at below-cost bargain prices. Marketing agreements made producer participation obligatory, leading to product standards that contributed as well to soil and water conservation.[43]

Following passage of federal marketing legislation in 1933, county agents in the San Luis Valley brought together growers and shippers to create Colorado's first formal marketing agreements, covering cauliflower and peas. After passage of the Colorado legislation in 1939, county agents facilitated the drawing up of marketing agreements starting with peaches, onions, and potatoes. The process was entirely democratic: the Colorado Department of Agriculture scheduled a referendum in which all producers of a given crop were eligible to vote. If a predefined majority of producers voted for self-assessment in the affirmative, then a state board of control was set up to receive and disburse revenues that by law were reserved for sales promotion, education, and research on a particular crop. During harvest time, a period of surplus, the agreements allowed for restricting shipments to higher grades that commanded better prices. By withholding lower grades, producers could actually earn more income than if they shipped all grades at reduced prices. According to state marketing experts, a 10 percent infusion of lower-grade product into the market could result in an overall 40 percent to 50 percent reduction in prices paid to producers. Extension horticulturist William M. Case expressed the hope that grading would encourage cultivation of the highest-quality crops over the long term. He recommended that assessment revenues be spent only on those crops for which marketing agreements guaranteed shipment of high-quality produce, crops for which Colorado could claim a special niche. Colorado cauliflower dominated US rail shipments during August and September; Western Slope peaches occupied a particular niche in a well-defined truck transport territory. Colorado produced the greatest amount of pinto beans of any US state, best known for their exceptionally high Vitamin B content.[44]

Similarly, county agents in the San Luis Valley assisted the Potato Growers Exchange, still headquartered in the valley, in establishing the Colorado Potato Administrative Committee as the board of control to administer the potato marketing agreement statewide. Dividing the state into three production areas—northern Colorado, western Colorado, and the San Luis Valley—the committee had authority to restrict shipments of lower-quality potatoes from any or all areas; it allowed its area subcommittees to further regulate shipments by grade, size, and date of maturity and to devise their own promotional campaigns. The committee did require all growers to use standardized shipping containers labeled according to the grade of their contents.[45]

Despite the many state and federal support programs, Colorado's farm economy remained in decline. Between 1930 and 1940, sales of farm products fell by one-third, despite bumper harvests in 1938–39. The number of farms fell by nearly 10 percent, to 59,956, but average farm size increased by nearly 25 percent, from 482 acres to 613 acres. Land under cultivation expanded by nearly 9 percent, to 31.5 million acres, but irrigated cropland declined by 25 percent, to 2.5 million acres. Full-time employment shrank significantly; nearly one-third of farm owner-operators held part-time jobs off the farm, especially along the Front Range where off-farm work was more lucrative than farming.[46]

No place better illustrates changes in the use of the land from rural to urban, agricultural to non-agricultural, than Jefferson County, Colorado. Whether prompted by feelings of nostalgia or exhilarated by the inevitability of growth and development, the present-day traveler on Interstate 70 from Denver west passes shopping centers, business parks, and suburban neighborhoods before reaching the foothills dotted with high-end homes and parkland. It is worth considering how that transformation occurred. Recall that the earliest Front Range farmsteads were located along Clear Creek, supplying produce to miners in the nearby mountains. With the decline in mining and the arrival of the railroad connecting Golden to Denver in 1870, farmers began shipping grains, fruits, and vegetables to the city by rail; ranchers still trailed their livestock to Denver stockyards into the 1920s. The county's population was less than 20,000 in 1917 when the extension service placed its first agent in Golden. Roughly one-fourth of the county's land surface was under cultivation: about 2,500 acres in vegetables, 350 acres in small fruits, 750 acres in orchard fruits, and 200,000 acres in commodity crops and hay. Ranchers grazed livestock on the thousands of acres of National Forest lands in the mountainous western part of the county.[47]

With approximately 2,000 farms, Jefferson County in the late 1930s still appeared agricultural, although the majority of its population was considered urban because of its proximity to Denver. Increases in average farm size had obscured the fact that a few farms furthest from the city had consolidated

FIGURE 4.8. Denver Union Stockyards, 1915. *Courtesy*, Denver Public Library, Western History Collection, MCC-4205.

while the majority of farms closer to the city operated on small, intensively cultivated tracts, of which 40 percent were 10 acres or less. In 1940 Jefferson County ranked among the top three Colorado counties for the production of asparagus, green and yellow beans, carrots, celery, sweet corn, cucumbers, summer and winter squash, and tomatoes. The only large industry in the county was Coors Brewery, taking advantage of Clear Creek for water and local farms for grains.[48]

Jefferson County agents continued in their traditional role as agriculture advisers, advocating for careful cultivation as the surest, safest, and least expensive approach to high yields and profitability. Agent Louis G. Davis had gone so far as to recommend biological control of low to medium infestations of harmful insects, based on his observations of large groups of different species of ladybugs feeding on aphids. His successor, Charles M. Drage, reported on his efforts to persuade farmers that insect populations would remain endemic and not become epidemic if soil health were maintained. Again and again, agents expressed their frustrations with farmer apathy toward scientific agriculture and reluctance to band together in their own self-interest. To counter chain grocers that were filling local stores with produce from the West Coast, Davis brought in state marketing director Ben King to lay out the structure for an association of Jefferson County market gardeners; after several attempts, the venture failed to attract sufficient grower commitment. Agent Davis reported that "every man is afraid of the other fellow and none of them will trust anyone who might be secured as manager of such a concern."[49]

By the late 1930s, the encroachment of suburbia into Jefferson County made land more valuable for real estate development than for farming and ranching. In an attempt to manage suburban development, representatives from local governments and businesses in a five-county area created a regional planning council in 1938, reorganized five years later as the Denver Regional Planning Association and now the Denver Regional Council of Governments. With the influx of new residents seeking urban amenities such as parks and safe, attractive neighborhoods, legislators representing the suburban counties successfully introduced legislation in 1939 that granted counties authority over planning and zoning in areas outside municipalities. This complemented legislation passed ten years earlier that enabled municipalities to conduct land-use planning. Jefferson County was the first to establish such regulations, designating "A-1 Agricultural" as the zone to protect the few remaining farms, with no restrictions except for compliance with state health regulations regarding domestic water and sewage treatment. Setting aside county land for agricultural preservation would come later.

In 1940 the US Department of War, as part of wartime mobilization, purchased the 2,040-acre Hayden Ranch in Jefferson County as the location of an ammunition factory and contracted with the Remington Arms Company to build the Denver Ordnance Plant. At peak production, the plant employed 22,000 workers. After the war the plant was dismantled, and the land was redeveloped as site of the Denver Federal Center, one of several permanent federal installations in the county. The ensuing demand for housing contributed to the complete conversion of the plains portion of the county to nonagricultural uses.[50]

In the county foothills, meanwhile, the Soil Conservation Service had secured the cooperation of landowners to participate in soil restoration projects. Of particular note, by taking the 9,000-acre Ken Caryl Ranch out of bankruptcy in 1937, the National Surety Insurance Company of New York agreed to invest in reclamation of overgrazed areas and washed-out gullies.

After mapping soil conditions on the ranch, the SCS drew up plans to build a series of small temporary catchment dams, releasing water as needed into newly reseeded pastureland; install fencing to protect eroded areas; and initiate rotational grazing. In addition, the SCS and extension sponsored range improvement demonstration projects on the Hiwan Ranch, located in the more mountainous portion of the county around present-day Evergreen. In 1942 county residents voted to tax themselves for support of a conservation district covering more than 150,000 acres. No question, urban and suburban dwellers were taking interest in preserving open spaces, although not so much for agricultural purposes.[51]

As the nation became more urbanized, the more progressive church leaders sought ways to connect and in some cases to reconnect rural congregations to the land, celebrating what might be described as the moral imperative of faithful stewardship—very much in the spirit of Liberty Hyde Bailey's *The Holy Earth*. Secretary of Agriculture Henry A. Wallace, an ordained Presbyterian minister, embraced that approach. His denomination had joined the campaign initiated and sponsored by the Federal Council of Churches (now National Council of Churches), emphasizing the moral aspect of agriculture as a means to encourage spiritual revival among rural congregations. Since 1929, the council had urged member denominations to set aside the fifth Sunday after Easter as Rural Life Sunday—a sort of pre-Earth Day—to celebrate their connection with the land; this was quite popular during the 1930s but became moribund during the war years. Afterward, the National Association of Conservation Districts sought to revive the celebration as National Soil and Water Stewardship Week, held the last Sunday in April through the first Sunday in May. In 1954 the Colorado Association of Soil Conservation Districts took up the cause but failed to move its own membership, engaging very few pastors. Parallel in time to the Protestant national council but not much more successful, the Roman Catholics had established the Catholic Rural Life Conference by seeking to revive and modernize Rogation Days, a period of fasting, prayer, and procession tied to spring planting; they celebrated Rural Life Sunday on the fifth Sunday after Easter.[52]

By the late 1930s, the principles and practices of federal financial and technical support had become institutionalized. They were generally accepted even by those agriculturists who were the most outspoken against government. This was made easier after President Roosevelt committed to sell agricultural products to Great Britain before America formally entered World War II. Once again, the USDA emphasized maximum production, exercised more flexibility in enforcing rules, reduced penalties on farmers for exceeding acreage crop allotments, lifted marketing quotas, and more broadly defined soil-depleting and soil-conserving crops to preserve conservation payments to farmers. While those programs benefited large operators the most, the Farm Security Administration made micro-loans available to small-scale farmers, even though some argued that keeping such farmers in business was highly inefficient. In July 1941, Claude R. Wickard, the new secretary of agriculture, established state and county defense boards (later known as war boards) to adjust wartime emergency regulations, redirecting some USDA field activities. As with earlier emergencies, the secretary designated county agents as executives of the local boards, explaining to farmers the various government programs and how to access them to meet production goals.[53]

County war boards reviewed and acted upon applications by farmers seeking exemptions from wartime rationing of items such as petroleum products and tires and collaborated with military draft boards in the matter of agricultural deferments. County agents assisted farmers in obtaining and often sharing equipment. Perhaps most critical, war boards participated in the recruitment and placement of farm laborers. Following the bombing of Pearl Harbor, when many young farmers enlisted or were drafted into the military, the Colorado war boards engaged in the recruitment of temporary labor, mostly migrants from New Mexico and, beginning in 1943, prisoners of war through Camp Carson near Colorado Springs. The farm bureau had successfully lobbied Congress to prevent the USDA from setting and mandating minimal wage, hour, and housing standards for migrant workers; county agents reported on their poor living conditions in jerry-built labor camps. In contrast, prisoners of war (POWs) were well treated as required by international conventions. Toward the end of the war, the military allowed POWs to work on farms without guards. It was no secret that famers much preferred POWs as laborers, and as many as 500 per month were placed in Colorado agricultural jobs.[54]

To complement work on the farms, demonstration agents, by then retitled home economists, instructed householders on using substitutes for foods unavailable or in short supply. In an era before home freezers, they demonstrated preservation techniques—salting, pickling, dehydration, and canning. Jefferson County agent Charles Drage opened the first extension office in Denver County; home economists and 4-H leaders helped recruit and train volunteers to assist in the development and planting of Victory Gardens, a project of home gardens first instituted during World War I to help provide sustenance for families who could not afford to purchase canned foods. By the summer of 1944, there were 42,000 Victory Gardens in Denver County. At the time, about one-third of all produce nationwide came from Victory Gardens.[55]

Cultivating Victory Gardens was a wartime necessity for some and an opportunity for others to reacquaint themselves with the pleasures of cultivating and harvesting fresh vegetables and small fruits. But wartime brought a darker side. In the interest of national security, in retrospect deemed excessive, President Roosevelt ordered that West Coast residents of Japanese descent be sent inland to internment camps. Among the camps was the Amache Relocation Center, which operated from October 1942 to October 1945. Located in Prowers County on 11,000 acres of marginal farmland purchased by the federal government as part of New Deal relief and recovery, the camp barracks and mess halls accommodated a population that reached a peak of 7,318. Despite the stresses and strains of sudden detention, internees quickly turned the camp into a real community. They repaired neglected irrigation works and restored and cultivated surrounding government land. Many internees were experienced market gardeners. For their sustenance, they grew some produce not previously cultivated in southeast Colorado: varieties of celery, spinach, head lettuce, potatoes, lima beans, onions, mung beans, daikon, and herbs for teas. They opened a cannery to ensure the availability of produce year-round, cultivated grain crops, and operated their own livestock processing plant. Students at the camp high school developed and operated their own experimental farm, adapting Japanese intensive farming techniques to local conditions and organizing a chapter of Future Farmers of America. Internees cultivated the camp's sugar beet fields, about 9,000 acres, for the American Beet Sugar Company of Rocky Ford and hired themselves out to nearby farms as well. When the camp closed, roughly one-third of the internees settled permanently in Colorado, mostly in northeast Colorado and the San Luis Valley where, decades earlier, agriculturists of Japanese descent had established successful commercial farms.[56]

FIGURE 4.9. Amache camp taken from guard tower, 1942–45. *Courtesy*, Colorado State Archives, Denver, #60432, sub-box A, ff. 1, 4815.

Special wartime responsibilities placed on the agricultural college still allowed the experiment station to pursue its central mission of applying the results of research on campus to the state's farms and ranches. The station's most publicized testing at that time concerned dichloro-diphenyl-trichloroethane (DDT), soon to be the first widely used synthetic insecticide. In the early 1940s, the USDA had supplied experiment stations with small amounts of DDT for testing on local conditions. Created by a German university student in 1874, it was not until 1939 that a chemist working for the Swiss firm J. R. Geigy (now Novartis) demonstrated the effectiveness of DDT in destroying the Colorado potato beetle. At first used by the military against insects that carried diseases such as typhus and malaria, the insecticide was commercialized after 1945, compounded with other substances, and sold under a variety of trade names. Widespread public interest obliged the extension service to issue instructions on the uses of this "wonder insecticide." Extension specialist Charles Drage and entomology professor George M. List warned of DDT's highly toxic effects on fish and other cold-blooded creatures, with less immediate but cumulatively just as poisonous effects on warm-blooded animals. By injuring or destroying populations of bees and other beneficial insects, they had observed that DDT reduced pollination and increased numbers of harmful insects. In other words, DDT was no cure-all. Drage and List pointed out that natural insecticides such as pyrethrum and rotenone acted more quickly without DDT's residual side effects caused by exposure to light, wind, and other factors. Despite their warnings, growing use of insecticides and herbicides compelled extension agents to sponsor demonstrations on best ways to apply those products. In May 1945 the first application demonstrations, both ground spraying and crop dusting from the air, took place on Arkansas Valley farms. For large commercial operations that used fixed irrigation systems, aerial crop dusting soon became the norm.[57]

Reflecting on the effects of wartime mobilization on farmers in the San Luis Valley, county agent Albert Goodman lamented that "high prices for cash crops during war years caused farmers to increase their acreages of such crops to the detriment of good sound farming practices." Farmers had plowed up pastures to grow potatoes, which led to immediate postwar overproduction and more soil deterioration and upset the traditional farmstead balance of crops and livestock. With fewer fields for forage, farmers used their federal payments to buy hay in neighboring counties—in Goodman's opinion, an unsupportable practice for both farmers and the government. For the postwar era, Goodman envisioned "a long-time farming program" to restore and maintain soil fertility. By no means original, its key ingredients included federal loans and payments consistent with sound farming practices; certified seeds for higher yields and better-quality produce; labor-saving devices on the farm and in the home; attention to the health of family members, laborers, livestock, and crops; suitable housing and training for farm laborers; and growing both crops and livestock to attract and maintain year-round permanent labor. Goodman had long dealt with the consistencies and contradictions of New Deal agricultural policy. His proposal sought to capture the best and most applicable aspects.[58]

Although obscured by the urgencies of economic relief and recovery and the necessities of wartime mobilization, New Deal reformers still held to the agrarian ideal of the independent farmer, the family farm as essential to the nation's economy, and the rural community as the "nursery of steady citizens." In April 1949 Charles F. Brannan, the Denver native appointed secretary of agriculture by President Harry S. Truman, presented to Congress a set of proposals by which the government would guarantee the preservation of that agrarian ideal: federal support of large-scale farm operations limited to the amount an efficient family-sized farm could produce and a requirement that all farmers receiving federal support comply with government regulations concerning soil and water conservation. Recall that both Brannan and James G. Patton, by then president of the National Farmers Union, were Costigan protégés, which made it easy for their adversaries to assume that Brannan's proposals were Patton's handiwork. The Brannan Plan, as it became known, represented the high water mark of federal engagement in agriculture, the closest Congress would ever get to guaranteeing farm income and ensuring the survival of the much-revered family farm. By convincing a coalition of Republicans and conservative Democrats to oppose the plan, farm bureau lobbyists cleared the way for industrialized agriculture.[59]

5

Advances in Productivity

Delta County is considered Colorado's most diversified agricultural county, producing fruits, vegetables, grains, livestock, and even timber. In 1959 Delta County agent Carl H. Powell reported that agriculture was undergoing "tremendous change"; public agencies such as extension needed to keep up with those changes for their work to remain useful. A veteran of the extension service when its principal task was to administer federal relief and recovery, Powell and his fellow county agents found themselves under a new directive: to assist farmers and ranchers in making the land produce more and in marketing more efficiently. For counties on the Western Slope, distances and the cost of transportation to urban markets posed perhaps the most serious marketing obstacle.[1]

To be sure, yields per acre had been increasing steadily since World War I, but the rate of increased productivity accelerated dramatically during and after World War II, in great part due to technical innovations adopted to meet heightened demand for food and fiber; this contributed to the industrialization of agriculture. These innovations fell into three broad categories: new

https://doi.org/10.5876/9781646422050.c005

plant cultivars and livestock breeds, advances in mechanization supported by increasingly sophisticated computerization, and adoption of synthetic fertilizers, herbicides, and insecticides. The application of those innovations enabled farmers and ranchers to push productivity beyond the natural limits of the earth, understandable within the philosophical framework of the infinite perfectibility of humanity and its economic iteration: development without limits. At the time, only a few scientists and theologians warned of possible long-term negative repercussions. One is reminded of the Icarus legend: warned by his godly father against flying too low over the sea or too high near the sun for fear of destroying his wings of feathers and wax, in supreme self-confidence he flew too close to the sun, fell into the sea, and drowned.

Between 1945 and 1974, the value of Colorado agricultural products sold increased by two-thirds, from $179 million to $265 million, while the number of farms and ranches declined by almost half, from 47,618 to 25,501; average farm size increased by nearly two-thirds, from 761 acres to 1,284 acres. Production of corn and wheat, by then the state's leading agricultural commodities, increased by 400 percent, from 12.2 million bushels to 50.2 million bushels and 17.5 million bushels to 70.1 million bushels, respectively, although their annual value increased very little. Livestock followed a similar pattern, with greater production, higher revenues, and fewer but larger operations. The number of cattle nearly doubled during that period, from 1.8 million to 3 million head, and their value more than quadrupled, from $110 million to $480 million; the number of cattle operations declined by more than half, from 38,538 to 16,739. Hog numbers increased by about 10 percent; their value more than doubled, from $5.2 million to $12.4 million; and the number of hog farms declined by 700 percent, from 21,477 to 3,013. Although sheep numbers fell from 2.4 million to 985,000, sheep values rose from $24 million to $37 million; the number of sheep operations declined from 5,582 to 1,876. With regard to fruits and vegetables, the production and value of apples harvested dropped by 16 percent, far less than the decline in the number of apple orchards, from 7,811 to 1,255; the value of vegetable produce sold increased by 16 percent, from $10.7 million to $12.4 million, while the number of vegetable farms declined from 5,488 to 1,125 and their total acreage dropped nearly 250 percent, from 76,443 acres to 31,375 acres.[2]

In the immediate aftermath of the war, agricultural prices declined and costs of production increased; but this time farmers and ranchers could rely on many forms of federal financial assistance. The US Congress continued price supports at 90 percent of parity for wheat and corn, lifted all restraints on production during the Korean War, and, beginning with the Eisenhower administration (1953–61), allowed for flexible price supports and reductions in parity consistent with the political outlook of successive national administrations. President Dwight D. Eisenhower's secretary of agriculture, Ezra Taft Benson, opposed federal constraints on large-scale farms, arguing that the best way for farmers to succeed was to become more competitive both within and beyond the agricultural sector. He strongly supported federal funding for research in the areas of production, processing, and marketing of agricultural products.[3]

The Eisenhower administration inherited commodity surpluses, thereby facing the continuing challenge of balancing supply and demand. Rather than controlling production through acreage allotments and marketing quotas, Secretary Benson favored reducing price supports on surplus commodities. The Agricultural Act of 1956 authorized the US Department of Agriculture (USDA) to establish the "soil bank," paying farmers and ranchers for voluntarily taking land out of production and thus preserving their income while improving soil and water quality. The act authorized and appropriated monies for two programs. First, the Acreage Reserve Program sought to bring about an immediate reduction in the production of six basic crops including corn and wheat; the program lasted only three years because it was deemed too expensive and riddled with loopholes. Second, the Conservation Reserve Program (CRP) sought long-term changes from cropland to more conservative uses such as open range, pasture, and forest—thereby contributing to water quality, soil fertility, reduction in erosion, and wildlife habitat. Under the CRP, the federal government made rental payments annually for reserved land and contributed to the cost of its improvement.[4]

On August 7, 1956, Congress passed and President Eisenhower signed the act that established the Great Plains Conservation Program. During the relief and recovery period of the 1930s, the USDA's Great Plains Agricultural Council had begun discussing ways to provide assistance to farmers and ranchers over the long term: eliminating the need for emergency funds, stanching periodic out-migrations due to drought, and in general helping to preserve rural communities. As part of the Great Plains Conservation Program, the secretary of agriculture could enter into long-term contracts, up to ten years, with farmers and ranchers "to assist them in making orderly changes in their cropping systems and land uses which will conserve soil and water resources and preserve and enhance the agricultural stability of [the Great Plains states]." The program would take place in twenty-five Colorado counties, with principal responsibility for administering it resting with the Soil Conservation Service (SCS).[5] For the first time, the federal government contracted directly with individual farmers and ranchers. A broad variety of activities qualified for cost sharing including seeding cropland with perennials, reseeding rangeland, developing efficient irrigation for grazing land, planting windbreaks, and changing cultivation techniques on cropland. Participation in the program was voluntary and did not preclude recipients receiving assistance through other government programs. To draw attention to the program's emphasis on permanent change in agricultural techniques, monies unspent at the end of a fiscal year could be carried forward to future years.[6]

The Great Plains Conservation Program ended in 1981, but many agriculture-related programs conceived during the New Deal continue to the present time. For farmers and ranchers, arguably the most popular of these programs remained the Soil Conservation Service (now the Natural Resources Conservation Service), working through local self-governing soil conservation districts. By 1945, fifty-six separate districts had been organized in Colorado. That summer, representatives from thirty-seven of those districts met in Denver to form the Colorado Association of Soil Conservation Districts (CASCD); their immediate task was to join a multi-state delegation organized by Oklahoma governor Robert S. Kerr (D) and favored by Colorado governor John C. Vivian (R) to lobby Congress for legislation allowing conservation districts to bid on heavy soil-moving equipment declared as surplus by the US Department of War. That delegation, including three Colorado stock growers, received a friendly reception from President Harry S. Truman at the White House in January 1946; Congress acted favorably shortly thereafter.[7]

Building on that success, the association adopted a broad purpose statement in support of conservation: "to assist and cooperate with soil conservation districts of the State of Colorado, and all civic, farm, and livestock groups and other organizations and agencies, whether governmental or private, in the furtherance of the conservation and preservation of natural resources, the control of wind and water erosion and the development of sound land

FIGURE 5.1. Annual CASCD convention, 1951. *Courtesy, Agricultural and Natural Resources Archive, Colorado State University Libraries, Fort Collins.*

use."[8] Because the term *land use* has become controversial in some circles, it is worth noting that CASCD did not oppose public ownership and federal management of public lands. Several conservation districts had adopted land-use rules that applied specifically to rangelands; the intent, not entirely altruistic, was to end the seasonal movement of livestock so that measures could be implemented and enough time elapsed for the land to recover. Although the state supreme court declared those rules an unconstitutional restraint of trade, other CASCD-endorsed land-use regulations did survive.[9]

In southeastern Colorado, conservation district supervisors sought to hold landowners, especially absentee owners, responsible for controlling and preventing wind erosion on their respective properties. In 1951 the CASCD drafted and successfully lobbied for passage of the Colorado Soil Erosion–Dust Blowing Act. The act gave county commissioners the authority to investigate areas where property owners had allegedly failed to prevent wind erosion; consult with conservation district supervisors, landowners, and tenants; and contract for remediation if property owners declined to do so. Some districts sought to ban altogether the plowing of lands susceptible to wind erosion, but that effort failed to pass legal review because it was interpreted as an infringement on landowner rights.[10]

To reach the parts of the state that had no conservation districts and, more broadly, the general public, the association succeeded in attracting statewide newspaper and radio coverage. During CASCD's formative years (1948–54), the *Denver Post* and KLZ radio sponsored an annual contest for the best soil conservation practices. Each district board of supervisors could nominate three farms; winning farms and their sponsoring districts earned recognition in those media. The contest organizer, KLZ farm director and Colorado State College of Agriculture and Mechanic Arts (name changed in 1935) graduate Lowell H. Watts, became director of the extension service (1959–82).[11]

Despite reports of occasional tensions, the extension and soil conservation services complemented each other. State law actually mandated that directors of both agencies serve as ex-officio members of the state Soil Conservation Board, reorganized in the early 1950s as the State Conservation Board to advise the Colorado Department of Agriculture on the disbursement of state funds to conservation districts. The CASCD lobbied the legislature to require that a majority of State Conservation Board members be selected by conservation district supervisors rather than by representatives of government agencies as originally set forth by the legislature. Recall that county agents had worked hard for passage of referenda to establish the soil conservation districts; afterward, they advised district supervisors on how to implement soil conservation practices. Furthermore, the extension service through the 4-H engaged farm and ranch youths in conservation district projects, expecting that someday the former 4-Hers would adapt what they had learned to their own farms and meanwhile influence their parents to take up those practices.[12]

Beginning in the mid-1950s, statewide coordination of government-sponsored soil and water conservation projects took on some urgency; industrialization and urbanization brought new challenges for the agricultural sector, which represented a declining portion of the gross state product. At the time, eight different units of the USDA and twenty-two separate state agencies were involved in one or more aspects of conservation of the state's natural resources, with each agency operating independently on a matter that by its very nature required coordinated efforts. The CASCD took credit for convincing Governor Stephen L.R. McNichols (D) and the legislature to create the Department of Natural Resources, placing all conservation-related agencies under a single cabinet officer beginning in 1957.[13]

By then, drought of 1930s intensity had returned, this time affecting an area stretching from the Kansas state line west into El Paso and Pueblo Counties but causing less human stress and fewer financial losses, thanks to well-established state and federal assistance programs. Again, Baca County experienced the most severe dust storms and wind erosion. February 1954 seemed the cruelest month: winter wheat planted the prior fall, normally dormant, had winterkilled due to lack of moisture, leaving nothing to hold the soil—which had become pulverized by the normal sequence of freezing and thawing. Cyclonic wind storms caused widespread soil erosion. In June, intense heat and desiccating winds reduced the winter wheat crop to 5 percent of normal. July rains encouraged farmers to plant grain and cover crops, but extreme heat and drought recurred in August and September. Only with October rains did winter wheat have a chance to germinate.[14]

Since income from livestock represented the largest portion of the state's agricultural sector, it was no surprise that the CASCD focused its early efforts on the conservative use of rangelands. CASCD president George C. Elliott expressed the association's position: "All grass is flesh in the making . . . grass is a crop to be harvested." The purpose of range conservation, then, was to improve grasslands so as to increase carrying capacity within the limits of the land's capabilities, for which Elliott and his board developed an action agenda for the association: ensure that land users are represented in the development of land-use policies; advocate for long-term leases on public lands, which would encourage permittees to invest in range improvements; seek the elimination of burrowing rodents ("prairie dogs") and abandoned and wild horses; and encourage fencing to supplement natural barriers as lease boundaries. The CASCD adopted recommendations of the SCS, encouraging ranchers to locate salt blocks away from watering holes and other livestock concentration spots; build contour furrows, fill in gulleys, or construct barriers to control erosion; build more stock ponds to better distribute livestock; and make allowance for natural and, if necessary, artificial seeding along stream banks, whether intermittent or otherwise. The CASCD opposed leaving restoration of the land to nature alone. In effect, that would have eliminated livestock from rangelands, either limiting or cutting off a rancher's ability to earn a living as well as reducing local tax revenues. Failure to harvest grass "creates a fire trap in which food for livestock and game animals is entirely lost," Elliott wrote, and also "causes many thousands of dollars' worth of [losses] in property, and even human lives"—an argument still used by forest grazing permittees.[15]

Improving rangelands and improving livestock breeding methods went together, allowing ranchers to produce more beef and gain more income per animal. Since ancient times, farmers have deliberately bred animals to improve their stock, as we know from descriptions of jerry-rigged methods to collect and implant male semen. With the invention of the microscope in the seventeenth century, scientists begin to speculate further on the possibilities of artificial insemination. Commercial use, however, did not occur until the 1940s, starting with confined dairy herds. By the 1950s, scientists had learned how to preserve semen, so artificial insemination became possible with range cattle. Artificial insemination is considered among the earliest application of biotechnology, the manipulation of biological processes to improve reproduction and genetic makeup.

Animal and veterinary science faculty associated with the experiment station helped prepare county agents with information they could use to convince dairy farmers to overcome their hesitancy about using artificial insemination. Some of that hesitancy could be attributed to lack of knowledge and fear that artificial insemination led to abnormalities, but it also reflected reluctance to deal openly with anything having to do with sex. In the early 1950s, artificial insemination was administered to fewer than 10 percent of all dairy cows in Colorado. Recordkeeping on bull production, essential to the success of that process, had yet to be generally adopted.[16]

It is unclear exactly when artificial insemination of range cattle in Colorado came into use. In 1946 Colorado State College of Agriculture and Mechanic Arts (Colorado A&M) joined "Improvement of Beef Cattle through Breeding," an initiative supported by the USDA Bureau of Animal Industry involving experiment stations in twelve western states. Under the direction of Howard H. Stonaker, a pioneer in science-based cattle breeding, Colorado's participation took place using a herd of 150 purebred Hereford cattle on a 10,000-acre site owned and managed by Fort Lewis College in Hesperus near Durango. A former frontier army post, Fort Lewis had been turned over to the state for an agricultural and mechanical arts high school reserved for Ute tribal youths; it expanded into the two-year Fort Lewis Agricultural and Mechanical College and was then placed under the auspices of the State Board of Agriculture along with Colorado A&M, thus the research opportunity. Taking a cue from the recent hybridization of corn, researchers at Fort Lewis sought to discover whether crossing closely related (inbred) sires could produce hybrid Herefords. Because the herd was purebred, researchers had a clear base from which to test and compare inbreeding and out-breeding (mating unrelated or distantly related animals). To attract the interest of area ranchers, researchers explained their project in three steps: first, development of different inbred lines of Herefords, comparing weaning weight, grade, and efficiency of feed utilization; second, performance testing of prospective sires to determine efficiency of feed utilization and weight gain; third, crossing different inbred lines with performance-tested bulls, crossing performance-tested bulls with unrelated cows, and testing for hybrid vigor of the cows and their calves. At the annual college-sponsored auctions, ranchers who supplied stock for trials could purchase performance-tested bulls. At the end of the project's tenth year, researchers concluded that hybrid cows out-produced inbred cows by almost 43 percent in pounds of calves weaned, which they attributed to a 30 percent greater calf crop and 10 percent heavier weaned weight. Once artificial insemination became commercialized, cross-breeding would become far easier and less expensive to accomplish, resulting in greater efficiency and potentially higher profits for ranchers.[17]

More dramatic advances in breeding occurred with crops, vegetables, and fruits. As the prevailing commodity crop on the Great Plains, corn was the subject of the first commercially produced hybrid seed. Recall that Boulder County agent H. H. Simpson had introduced 'Minnesota No. 13' to local farmers and taught his corn club boys how to cultivate, identify, and select the best specimens for use as seed stock and keep yield records in successive years. But that was not hybridization. Credit for inventing or at least commercializing hybrid corn went to Henry A. Wallace, the future secretary of agriculture. As a young man he had experimented with cross-breeding, discovering that the first generation of a cross-bred plant was generally superior in quality to both parents. He applied his findings on hybrid vigor to Iowa's principal crop. In 1924 he began selling his 'Copper Cross' as the first commercially available seed corn. He then founded the Hi-Bred Corn Company (1926), which became Pioneer Hi-Bred Company (now DuPont Pioneer) and occupied the largest share of the nation's hybrid seed market.

In 1937 the Colorado Agricultural Experiment Station began testing corn hybrids at two sites. Researchers at Fort Collins used early maturing hybrids, comparing yields with 'Minnesota No. 13,' by then called 'Colorado No. 13.' Over a period of eight years, they discovered that hybrids yielded 19 percent more than 'Colorado No. 13.' Researchers at Rocky Ford tested hybrids against 'Reid Yellow Dent,' an open-pollinated later maturing cultivar, and recorded hybrids yielding 25 percent more than the standard cultivar. Similar to the beef cattle project, researchers made performance records available to corn growers and seed companies. In addition to yield of shelled corn per acre, records showed weight per measured bushel, percentages of moisture content at harvest, and presence of suckered plants (multiple stems), diseased plants, and broken stems.[18]

Since hybridization continues to play a significant role in all aspects of agriculture and is sometimes confused with genetic modification, it is worth pausing to consider how hybrids are developed using the example of corn, since we are all familiar with its ears and kernels. A corn plant is monoecious, containing both male and female sex organs on the same plant, and is self-pollinating. Under natural conditions, pollen grains fall from the tassel onto the sticky silks emerging from the top of the ear; upon reaching the silks, the pollen grains germinate, growing down the silks, with each silk attached to a single ovule. The fertilized ovules develop into seeds, the kernels we eat. In the greenhouse, the strategy is to prevent self-pollination in an effort to combine the pollen of one plant (the male parent) with the ovules of another (the female parent). Traditionally, the breeder accomplishes that by placing a bag over the ear, allowing it to continue to grow but preventing pollen from falling on the silk. When pollen begins to fall from another plant, the breeder fastens a bag over the tassel to trap the pollen. The next day, the breeder removes the bag holding the pollen, uncovers the ear, and quickly places the pollen on the silk. In the field, the strategy for cross-breeding is simpler: the breeder plants two inbred cultivated varieties in alternate rows, then removes the tassels of one variety, leaving only the female ears, so that pollen from the second variety—with help from the wind—fertilizes the first variety, producing hybrid seed.[19]

Hybridization is one method by which we help nature create more productive plants. But as Gregor Mendel discovered, nature limits hybrid vigor and consistency to the first generation; beyond that, the cross-bred plant does not remain "true" to the first generation and becomes variable. In practical terms, farmers using hybrids must purchase seeds annually from the seed companies that hold patents on those creations. The widespread use of hybrids comes with costs beyond seeds. Hybrid cultivars generally require more fertilizer and more water. In addition, their widespread use has certain societal implications. Small-scale farmers who cannot afford hybrid seeds and associated costs might be pressed to sell out to larger operators or even abandon their farmsteads and communities. As farms grew larger and more mechanized, growers benefited financially by cultivating a single cultivar. The resulting decline in biodiversity posed risks yet not entirely understood.

Field testing of hybrid cultivars of vegetables began in the 1950s. Colorado A&M partnered with other land-grant colleges in a USDA-sponsored national onion-breeding program. This program was of special interest to Colorado, ranked second in the nation for onion production. Colorado trials took place in all three principal onion-growing areas: the Western Slope near Olathe, the northeast plains, and the Arkansas Valley. Hybrid onions bred in those particular locations generally yielded more than long-standing cultivars. Using cultivars from other parts of the country, Colorado researchers were able to breed for resistance to "pink root," a disease caused by the fungus *Phoma terrestis*. Endemic to Colorado soils, "pink root" causes onion bulbs to become stunted, reducing yield and bulb size and making them unsalable. To encourage the production of high-quality seed for sale to seed houses, in 1949 extension horticulturists organized the Colorado Certified Onion Seed Growers Association, which later became part of the Colorado Seed Growers Association, still closely associated with the extension service.[20]

Complementing advances in breeding, rapid mechanization of agricultural production was made possible by wartime technological innovations, prompted by federal agricultural policy that encouraged efficiency. Foremost among the boosters of mechanization were sales representatives of agricultural implement companies and agricultural lenders, followed by small- and large-scale farmers who saw opportunities for higher profits. As a result,

FIGURE 5.2. Sugar beet factory, Fort Morgan, 1950. *Courtesy*, Agricultural and Natural Resources Archive, Colorado State University Libraries, Fort Collins.

increasingly specialized and sophisticated machinery impacted every nook and cranny of agricultural production. Annual conventions of the CASCD provided ideal venues for companies to display their new machinery and demonstrate how their products could help farmers and ranchers earn more while carrying out soil and water conservation projects.[21]

Delegates to CASD's 1967 annual convention may have experienced mild indigestion while listening to their banquet speaker talk about the "highly mechanized agricultural factory" taking shape in rural America. Russell R. Poynor, research director of International Harvester's farm equipment division, noted that mechanization had freed individuals traditionally tied to farming to move into non-agricultural jobs without jeopardizing the nation's food supply. Small-scale farm operations were no longer viable, so the farmer of the 1960s had to be "a combination of business executive, animal nutritionist, agronomist, soil scientist, economist, and engineer all in one package." Similarly, the basic tractor no longer sufficed, evolving into an ever more precise and versatile piece of machinery. Poyner explained that under a load, the hitch on the new International Harvester tractor automatically adjusted itself "three times a second in as fine a movement as 1/16th of an inch—and the operator doesn't have to do a thing to it." More technically advanced than the automobile, the tractor was equipped with power steering, power shift torque amplifier, power disk brakes, and power lift; automatic controls with multiple outlets for operation of implements from the tractor seat; sixteen forward and eight reverse speeds, and much more. With all those accessories and greatly increased power, the modern tractor made possible larger farms, greater productivity, and more leisure time for farmers. Coming innovations would reduce cultivation to two passes annually over a field—once for soil preparation, seeding, and application of chemicals and once for harvesting. Poynor correctly anticipated the day when the harvesting and processing of crops could be completely mechanized, controlled, and monitored by on-tractor computers. As to the soil itself, he speculated that it might merely serve as "a means to physically anchor the crop, while it is exposed to the atmosphere and sunlight for photosynthesis." At that point, he added, soil should no longer be called soil; presumably, it is just a receptacle of production.[22]

The widespread adoption of commercial fertilizers proved as significant a factor in increasing yields while keeping food costs low as was the development of new breeds and advances in mechanization. In 1954, the first time the US Census recorded its use, an estimated 11 percent of Colorado farms used commercial fertilizer, at an annual cost of $700,000; in 1969, an estimated 38 percent of farms did so, at a cost of $20 million. At the same time, the value of all crops harvested annually rose from $108 million to $341 million and in 1974 to $955 million. By 1969, anhydrous ammonia in liquid form was the fertilizer of choice. Its high nitrogen content of 82 percent compared favorably to ammonium nitrate, the more common postwar commercial fertilizer marketed in solid form, which was less convenient to apply and quicker to leach through topsoil.[23]

The use of commercial fertilizers accelerated efforts to more precisely understand the nature of soil fertility. In a short piece titled "The Use of Commercial Fertilizers," written in 1946, Delta County agent A. F. Hoffman reflected on how decades of harvested crops had extracted vital ingredients

FIGURE 5.3. Tractor with planter, 1969. *Courtesy*, Agricultural and Natural Resources Archive, Colorado State University Libraries, Fort Collins.

from the soil, erosion had washed out minerals, and irrigation had changed the very nature of those soils—resulting in lower yields and poorer-quality produce. He reported that soil scientists and those who study plant food had yet to agree on the number of elements plants needed to flourish. For a long time, scientists believed that only four elements were required: calcium (lime), nitrogen, phosphorus, and potassium. By 1946, scientists supposed that sixteen elements, divided into four groups, were needed: (1) carbon, hydrogen, and oxygen from the air and water; (2) nitrogen, phosphorus, and potassium, known as the primary elements; (3) calcium, magnesium, and sulphur, the secondary elements, often referred to as soil amendments; and (4) trace elements including boron, iron, manganese, zinc, copper, chlorine, and molybdenum. In noting the limitation of research to date—his list has since been revised—Hoffman meant to caution his constituents about the use of commercial fertilizers. His advice: learn what elements a specific plot of soil might lack, what a particular crop would need, the preferred method for spreading fertilizer, how much should be applied per acre, and the best time to do so. "The purpose of commercial fertilizer," he cautioned, "is to replace plant food the soil has lost and make useable that which is still in the soil." As part of an overall program of farm management, commercial fertilizer helps repair the damage to nature caused by cultivation, the implication being that commercial fertilizer cannot be used to improve upon nature. In contrast, chemical company representatives rightly or wrongly believed their products would overcome the limits of nature, ensuring sufficient food to feed the world's growing population indefinitely.[24]

With agriculturists increasingly relying on synthetic fertilizers, county agents were no longer perceived as the single source of agricultural knowledge. The extension service did retain its unique role of providing farmers and ranchers with objective data, ranging from seed testing to soil and water analysis to diagnoses of plant and animal diseases. Indeed, agricultural chemical dealerships and lending institutions supported extension's soil testing and analyses. Dealerships provided test kits to farmers at no cost; banks helped pay for the tests, expecting farmers to borrow more so they could purchase those fertilizers. As with any new business, the chemical industry had its share of unscrupulous agents. County agents were in a position to monitor, even help, farmers seek redress for their complaints. In Weld County, a business calling itself Standard Fertilizer Company had sold thousands of dollars worth of a product that proved totally ineffective. Agent George L. James assisted farmers in obtaining a restraining order from the Colorado Department of Agriculture against further sales by the company in their county.[25]

Synthetic fertilizers had proven far more effective on irrigated crops than on non-irrigated crops. But in 1952, scientists at the Akron research station began experimenting with adding nitrogen and phosphorus to dry-land farmed fields of corn, winter wheat, and grain sorghums. Their efforts with water conservation, better weed control, and stubble mulching had improved winter wheat yields; but protein content had declined as the soil's natural nitrogen content did not regenerate quickly enough to keep up with crop requirements. By adding nitrogen, scientists did observe some gain in yields on sandy soils but less on clayey hardpan. The challenge was to apply fertilizer at just the right time, covering moist soils when fields lay fallow.[26]

FIGURE 5.4. Flood irrigation, Longmont, 1963. *Courtesy*, Agricultural and Natural Resources Archive, Colorado State University Libraries, Fort Collins.

As with inorganic minerals used for fertilizer, natural herbicides and insecticides had long been used by agriculturists. Recall that early Greeley-area farmers used Paris green (arsenic) against potato beetles; Bordeaux mix—copper sulfate and lime—helped protect cantaloupe vines in the Arkansas Valley. Plant-based rotenone, also developed in France but recently banned, had been used as a pesticide by farmers against cabbage worms and other leaf-eating larvae. In addition, the practice of rearing "good bugs" in captivity and then releasing them in orchards to control "bad bugs" had been known for centuries. In the early 1940s, the Colorado Department of Agriculture had established the Palisade insectary, which over time became a significant source of information for agriculturists seeking to reduce their reliance on chemicals. Strategically located in orchard country, the insectary began by growing the plants for and then raising the insects that feed on pests such as the fruit moth, expanding to insects that feed on noxious weeds such as leafy spurge and field bindweed. To date, biological products alone remain insufficient in supply and too costly to control weeds and pests on the more than 36 million acres of the state's cropland and rangeland.

DDT had emerged postwar as the insecticide of choice, although quantities were limited and costs remained relatively high. As a result, to cope with grasshopper infestations in Baca County, agent James E. Hughs spent more than a month in the spring of 1949 operating a mixing station. He reported using 20,000 pounds of sodium fluorosilicate, one railcar load of bran, and three loads of sawdust. Great care had to be taken in handling the white granular powder, which irritated eyes on contact and caused breathing problems when ingested. Ranchers whose properties adjoined the national grasslands asked Hughs to intervene on their behalf by requesting that federal officials bait the grasslands. Government officials declined the appeal on the grounds that the offending grasshoppers were identified as native; the density of 10 to 15 grasshoppers per square yard was less that the officially set epidemic level of 24 or more per square yard. Officials did, however, assign personnel and spray equipment to areas invaded by migrant grasshoppers at the epidemic level. As drought conditions worsened, epidemic levels reached up to 300 grasshoppers per square yard in Cheyenne County, which became the site of the first massive aerial spraying of DDT. Decommissioned four-engine B-17 bombers conducted early morning runs over 271,000 acres. In 1958 state and federal agencies collaborated with local agencies and ranchers in spraying more than 3 million acres statewide, considered the largest spraying project in the world to that date.[27]

Just as DDT had become the preferred choice of insecticide, dichlorophenoxyacetic acid (2, 4-D) became the leading herbicide, introduced commercially in 1945 by the Dow Chemical Company. Known as a synthetic auxin, 2, 4-D is a chemical compound that causes plant cells to divide and

FIGURE 5.5. Sagebrush eradication test plot, 1951. *Courtesy*, Agricultural and Natural Resources Archive, Colorado State University Libraries, Fort Collins.

grow without stopping, burning more energy than they can make by photosynthesis and killing plants by over-stimulating them. Most useful from the perspective of farmers and ranchers, 2, 4-D destroys broadleaf plants but not grasses, which allows for aerial spraying onto fields planted in wheat and corn as well as on rangelands where, by eliminating sagebrush and sowing grasses, ranchers sought to increase livestock-carrying capacities.

The popularity of 2, 4-D among farmers and ranchers caused experiment station staff to further study its benefits and limits, field testing for best uses. As the lead state official overseeing weed control, seed laboratory manager Jay Thornton was convinced "that the discovery of the herbicidal qualities of 2, 4-D is the greatest single contribution to weed control efforts in the history of agriculture." But he also warned that "the extravagant claims made in over-zealous and premature reports have led the public to expect the impossible." Those claims were likely part of marketing efforts aimed primarily at homeowners seeking quick and easy ways to eliminate broadleaf weeds in their lawns without harming the turf grass.[28] An exception was perhaps Boulder County, where in early 1951 homeowners speculated that widespread damage to their flowers and gardens had been caused by disease. Extension staff had determined that the culprit was herbicide drift. County agent Edwin B. Colette reported that 2, 4-D aerial spraying over wheat fields southwest of Longmont had drifted over an adjoining cut-flower farm, destroying its entire crop. Colette noted that drift occurred even after the herbicide vaporized, forming a cloud close to the ground; the destruction was also caused by sprayed fields leaching into groundwater and affecting nearby untreated plots.[29]

Since the 1920s, residents in several Colorado counties had voted to establish weed and pest control districts, which made sprayer machinery and supplies available to farmers, ranchers, and railroad and irrigation companies; district staff sought to control weeds on road rights-of-way. With the postwar boom in the construction of new roads and reconstruction of established state and federal roadways, soil disturbance on rights-of-way brought in noxious weeds that competed with native plants. Not until the 1960s did highway departments write into construction contracts the requirement of reseeding, most commonly done with naturalized crested wheat grass (*Agropyron cristatum* [L.] Gaertn.). By then, county agents such as Weld's George L. James worried that weeds "spreading at an alarming rate" compelled farmers to urgently gain "a thorough knowledge of their causes and the application of this knowledge to farming practices." To be sure, the presence of weeds is as old as agriculture itself; farmers had become so accustomed to them that they were taken for granted. In so doing, farmers allowed weeds to exact enormous financial tolls, directly and indirectly—deflecting water, sunlight, and nutrients from crops; damaging produce; and increasing costs of cultivation and irrigation. Extension agronomists estimated that nearly one-third of all cultivated land, more than 9 million acres, was infested with weeds. Farmers spent over $20 million annually on weed control and still lost more than double the value of the crops harvested.[30]

As various forms of urbanization took place, soil disturbances invited more and more weed infestations on lawns and gardens as well as on agricultural lands. In response, chemical companies developed and marketed

FIGURE 5.6. Herbicide experiment, 1965. *Courtesy*, Agricultural and Natural Resources Archive, Colorado State University Libraries, Fort Collins.

an ever-increasing number of new herbicides—some tailored to specific soil and climate conditions, others to eradicate noxious weeds such as the ubiquitous perennials, Canada thistle and field bindweed. In 1965 the chemical industry sponsored the first statewide agricultural chemicals exposition, featuring a presentation "Weeds, an Expensive Luxury" by a representative from the Monsanto Company of St. Louis. Five years later, a Monsanto chemist discovered the chemical compound glyphosate and its value as a broad-spectrum herbicide. Introduced commercially in 1974 as Roundup, glyphosate became the world's best-selling herbicide, in recent years the subject of considerable controversy.[31]

Again to be clear, weed is not a botanical term; there is no single family, genus, or species of weed. In the words of Colorado weed scientist Robert L. Zimdahl, a weed is simply "a plant growing where it is not desired." Traditionally, the practical concern has been to control, if not eradicate, weeds. But in the early twenty-first century, Zimdahl was among the first scientists to ask some truly fundamental questions: What is it about the ways we treat the soil that allows weeds to flourish, and how might we change our ways of cultivation to reduce the costs of weed control while acting as good stewards of the earth—very much in the spirit of Liberty Hyde Bailey's *The Holy Earth*. Zimdahl concluded his distinguished tenure at Colorado State University as combination weed scientist and agricultural ethicist.[32]

Some farmers and ranchers expected synthetic chemicals to eliminate weeds and pests once and for all time, but experiment station researchers and extension agents continued to advise caution on their usage. As early as the late 1940s, researchers had discovered the limits and potential hazards of insecticides. The chemical used to control western yellow blight, which infected tomato plants especially on the Western Slope, sickened but did not kill the white fly (*Trialeurodes vaporariorum*) that carried the disease. Rather than fly away in search of other non-tomato hosts, the fly remained to infect additional tomato plants. Until a more specific spray could be developed, researchers recommended that growers plant double rows instead of using insecticides. Elsewhere on the Western Slope, agent Carl H. Powell cautioned Delta County farmers against relying exclusively on chemical sprays; he recommended the less hazardous and more conservative approach, which included the use of certified seed, timely tilling, alternate rotation of deep- and shallow-rooted crops, temporary electric fences to keep livestock out of cultivated fields, and allowing cattle to graze along irrigation ditches.[33]

At times not obvious, it is worth reminding ourselves over and over again that water continues to be the controlling factor in the development of the arid West. For Colorado agriculture, this has meant the full appropriation of water, moving surface waters from areas of surplus across watershed boundaries into areas of deficit and, to a lesser extent, bringing up groundwater for use in areas with limited or no surface waters. Added to that challenge, recall

FIGURE 5.7. First Colorado River water reaches northeastern Colorado, 1956. *Courtesy*, Agricultural and Natural Resources Archive, Colorado State University, Fort Collins.

that between 70 percent and 80 percent of Colorado's water supply is west of the Continental Divide; between 80 percent and 90 percent of the state's population now lives east of the divide.

Although diverting water from its natural course is documented as far back as the Ancestral Puebloans, the first trans-basin diversion project in Colorado was completed in 1880: the Ewing Ditch across Tennessee Pass carried 2,400 acre-feet of water annually (1 acre-foot covers 1 acre of land area 1 foot deep) from a tributary of the Colorado River to a tributary of the Arkansas River. Built to supply farmers in southeastern Colorado with irrigation water, the ditch and its water rights were eventually sold to the City of Pueblo. Of the twenty-one trans-basin diversion projects under construction in 1946, the largest and most significant was the Colorado–Big Thompson (C-BT) Project. Sold to the US Congress as a source of supplementary water to overcome recurring drought, the project enabled agriculturists to increase the number of their irrigated acres. In addition, beginning in the 1960s, Front Range cities began purchasing C-BT water for domestic and industrial uses, although this practice was challenged repeatedly in court by irrigation companies owned by irrigators.[34]

To document the early impact of the C-BT on agriculture, the experiment station sponsored a survey of 150 northeastern Colorado farms, taken before the project became operational in 1956 and again in 1964. Researchers learned that overall, the project had augmented existing eastern slope supplies of irrigation water by one-third. With additional water, farmers were able to plant and cultivate their fields more intensively, switch to row crops that provided higher yields, and anticipate greater profits. Farms increased in size, with more irrigated and fewer dry-land acres—all of which contributed to demand for better seed stock, more mechanization, and more chemicals.[35]

Moving water from one watershed to another within the same sub-basin or from one basin to another impacted livestock as well as crop production. In northeast Colorado, production under irrigation of more alfalfa, sugar beets, grains, and hay contributed to the expansion of feedlots, helping to meet consumer preference for confined grain-fed over range grass-fed livestock. Beyond the reach of trans-basin diversion, dry-land farmers benefiting from low rates levied by rural electric co-ops began large-scale pumping of groundwater. Unlike artesian water that flows naturally to the surface, the vast majority of groundwater in eastern Colorado required high-lift turbine pumps.[36]

Perhaps as significant for agriculture as the invention of Cyrus McCormick's mechanical reaper was Frank Zybach's "self-propelled center pivot overhead sprinkler irrigation machine," invented in Colorado. Born out of necessity, both inventions were hobbled together on the farm, using parts from old machines, scrap iron, and available hardware. Simply put, Zybach's machine rotated in a circle around a base pipe structure in the center of a field, applying water uniformly over a large area with a minimum of human effort. A dry-land farmer from the Strasburg area 50 miles east of Denver, Zybach patented his invention in 1952. He partnered with an automobile dealer in his hometown of Columbus, Nebraska, to build the first commercial models and awarded exclusive license to Valley Manufacturing in the 1960s. As pivot systems were improved to efficiently apply water and water-soluble nutrients, they could be used over a range of soil, crop, and topographic conditions. Valley Manufacturing expanded into Valmont Industries, becoming a global leader in the manufacture of agricultural irrigation systems. Today, a single center-pivot system with a radius of a quarter of a mile can cover up

FIGURE 5.8. Sugar beet harvest, 1960s. *Courtesy*, Agricultural and Natural Resources Archive, Colorado State University Libraries, Fort Collins.

to 90 percent of the area of a square field of 320 acres, with add-ons to cover the spots missed by the pivot.[37]

Anyone who has looked out of an airplane window while flying over the Great Plains and the West has seen the innumerable circular impressions made by center-pivot irrigation. In the wake of Zybach's invention, the number of deep wells dug for widespread irrigation and the pumps for raising the volumes of water required by the pivot system have grown exponentially—alarming irrigation water users, county agents, and public officials. Experiment station researchers in partnership with the US Geological Survey have studied changes in water table levels beginning in 1929, when wells were limited to providing water for people, garden plots, and livestock and fewer than five irrigation pumping stations existed in all of eastern Colorado. As pivot irrigation systems became operational—supplied by an estimated 700 irrigation wells in 1959—farmers and ranchers in eastern Colorado began reporting unprecedented declines in aquifer water table levels. Similar concerns came from irrigators dependent on South Platte and

Arkansas River surface waters. They attributed unusually low stream flows to the proliferation of alluvial wells (dug in stream beds and on floodplains), suggesting an as yet fully unexplored connection among surface, alluvial, and aquifer waters. Missing were basic data on groundwater volume, groundwater movement, and groundwater recharge rates.[38]

Colorado was years behind California and other western states in instituting the management of groundwater. The Colorado Ground Water Law of 1957 for the first time required obtaining a permit from the state engineer's office before digging a new well, as well as registering all existing wells with the exception of certain domestic, stock-watering, and artesian wells. The law created the Colorado Ground Water Commission to study the parts of eastern Colorado with little or no surface waters, where an ever-increasing amount of pivot irrigation had been depleting aquifer levels. Issuing a well permit did not mean granting a water right; a well, like a ditch, was viewed simply as a means to divert water from its natural place. Only in 1965 did the Colorado legislature create the legal framework for the appropriation and management of groundwater. For administrative purposes, the Ground Water Management Act classified groundwaters into four categories: (1) groundwater that is tributary to a natural surface stream, (2) groundwater designated by the groundwater commission as having negligible impact on surface-flowing streams, (3) non-tributary groundwater outside of designated groundwater basins, and (4) non-tributary and not-non-tributary groundwater in the Denver Basin. In *Colorado Water Law for Non-Lawyers*, P. Andrew Jones and Tom Cech explain that in Colorado all water is deemed moving, which means that surface waters and tributary waters (category one) are subject to the doctrine of prior appropriation and its administration by the Office of the State Engineer. Groundwaters in categories two through four are exceptions to that doctrine. The groundwater commission divides category two waters into eight separate basins on the eastern plains, allocating and administering those waters for the beneficial use of overlying landowners and maintaining sufficient balance between use and recharge so the supply lasts for 100 years, presumably from the date of development. Groundwater in categories three and four—non-tributary groundwater outside the designated area and in the Denver Basin—can be used without regard to recharge. Groundwaters in categories two through four may be considered nonrenewable natural resources.[39]

The increasingly sophisticated management of water combined with technological advances in cultivation, supported by federal funding priorities to ensure consumers of plentiful food at low prices, accelerated the trend toward large agricultural businesses. In the early 1960s, the Gates Rubber Company of Denver was considered the first non-farm corporate owner of a large ranch operation: Stateline Ranch in North Park, from which several thousand head of beef cattle were shipped annually to feedlots in Nebraska and California. In the late 1960s, Gates became the first non-farm corporate owner of a large farm operation. Gates Farms, Inc. encompassed 6,000 acres of rangeland in Yuma County that the company converted into irrigated farmland. This land was dependent on groundwater for producing a variety of crops, some of which required great amounts of water. Although the Gates farm ultimately failed, the company had sparked strong opposition among local agriculturists who viewed large corporate farms as an existential threat not only to their own smaller operations but to the very existence of their small rural communities. That story is told elsewhere; suffice it to note that farmer opposition gave rise to statewide debate on the natural limits of the land and, for some, the issue of unfettered economic growth.[40]

In November 1968, addressing delegates to the Rocky Mountain Farmers Union convention, incoming president Charles Hanavan Jr. of Cheyenne Wells put the "bigger is not always better" argument within the context of recently passed federal environmental legislation, President Lyndon B. Johnson's "war on poverty," and growing consumer interest in healthful foods. A strong proponent of the agricultural sciences, Hanavan argued for "a sensible compromise between the top-heavy practice of pouring more tons of fertilizer and other chemicals on farm land, and the sounder method of crop rotation for disease and insect control." An agrarian realist, Hanavan stands out as the first agricultural leader in Colorado to make the case for small and mid-scale farmers and ranchers as the best stewards of the soil, the most efficient producers of the nation's food over the long term. By connecting conservation of land and water with the well-being of rural communities, Hanavan, too, picked up on the theme of the moral obligation to serve as stewards of the "holy earth."[41]

Earlier, Thomas E. Howard, retired as the first secretary-treasurer of the Colorado Farmers Union, had written a handbook for rural pastors published by the National Catholic Rural Life Conference. Howard prefaced his handbook with a polemic against industrialized agriculture as a major contributor to rural poverty. He was careful, however, to distinguish between industrialized and mechanized agriculture. "Machines can and should do the arduous toil of men on all farms," he wrote. "On the other hand, the industrialization of agriculture permits machines to plow under the farm families, the schools, the churches, [and] the country roads." In connecting the teachings of the church with farming and rural living, Howard expressed the opinion that human greed had blighted the land in the form of erosion and caused "scars of human erosion on the face of society."[42] The rural pastor, he continued, needed to do more than learn and moderate discussions about farm problems; the pastor must be prepared to assist farmers in making the changes necessary to address those problems. Howard suggested that the best vehicle for changes was "cooperative conservation" through associations of farmers and other rural stakeholders, including county and home demonstration agents. The pastor's foremost duty was "to help restore a love of the land to the hearts of the people in his parish."[43]

Since rural pastors generally did not record their sermons, we have no direct evidence that materials their church organizations provided to instill a sense of Christian stewardship were ever used. Had they been, their impact was minimal at best, judging from the success of industrialized agriculture and the decline of rural communities. An exception was a small group of Benedictine nuns who had left the Abbey of St. Walburga in Eichstätt, Germany, in 1935 to establish a convent in Boulder County. Stewardship of the soil was a practical necessity for this ancient order. The nuns took over 150 acres of abandoned farmland close to Boulder, restoring the land to productivity. As the city expanded, they moved further out in the county, developing a successful farm operation that conserved soil fertility while producing modest yields. In the late 1990s, the nuns exchanged their highly appreciated property for a ranch location more compatible with their contemplative life, northwest of Fort Collins near the Wyoming state line, where they raise grass-fed beef cattle.[44]

Not of the modern world, the nuns were living according to the rule of their fifth-century founder, St. Benedict, when there was no alternative to natural farming. By deliberate choice, former Texas rancher Thomas M. Lasater (1911–2001) applied the techniques of natural farming to raising beef cattle, creating for himself a unique niche in which to earn modest profits during difficult economic times. In the aftermath of the stock market crash in 1929, he had dropped out of Princeton University, returned to the family ranch, and centered his attention on raising livestock as inexpensively as possible by reducing costs of feed grains, machinery, and labor. By leaving to nature, as he put it, most of the cattle-raising work, he developed Beefmaster, a new breed that combined characteristics of Hereford, shorthorn, and Brahman cattle. Shunned by his more conventional stock-grower neighbors and after searching for more land, Lasater purchased the 30,000-acre Matheson Ranch (Elbert County) in 1948.[45]

Lasater told delegates at the 1968 Colorado Association of Soil Conservation Districts convention that by the time of his purchase, Matheson Ranch had been mistreated for a century: ground cover was in "horrible shape," hundreds of acres were barren of native short grass, and the ranch had lots of jackrabbits and cottontails and a few antelope but nothing else. Twenty years later, by leaving nature alone, wildlife had returned to almost perfect balance, and ground cover was totally restored. To the consternation of some of his listeners, Lasater noted that the ranch prohibited hunting, trapping, or poisoning of predators as well as poisoning of ants and vegetation, minimally using spray on cattle because "we don't want to keep all flies and lice off, but just keep them under control." To maintain between 1,000 and 1,200 head of cattle year-round, the ranch put up 98 miles of fences, thirty-five windmills, three electric pumps, and four sets of working pens; it also maintained one pickup and one old tractor with a posthole digger.[46]

What began as the result of a purely business decision turned into an effective counter to industrialized agriculture, a lesson on mimicking nature over the long term although not entirely; since livestock are domesticated animals, their numbers are controlled by humans. Lasater maintained what he called a "balance of nature." In the course of day-to-day operations, his two sole ranch hands kept two essential records: an inventory of livestock numbers taken twice each month, which allowed for monitoring the true carrying capacity of the land, and a register of bull weights. During the prior ten years, it had taken an average of 44 acres to provide forage to sustain the equivalent of one cow and her calf (animal unit month, or AUM), which for the year 1967 amounted to managing an average of 1,077 animals on 24,402 acres. Of purely personal interest, Lasater kept other records, for example, to learn whether it was better for the land to graze two 1,500-pound or three 1,000-pound animals or, as he put it, subject the land to eight hooves or twelve. In the final analysis, economics, not conservation, ruled the ranch. Acknowledging that farmers and ranchers generally adjust very slowly to change, Lasater mused that he had witnessed more change in his business in the past two years than in all previous years combined. He ascribed that change to evolving consumer tastes and values, the public's positive reception to grass-fed beef, and its strong reaction to warnings about the health hazards of chemical additives used in industrialized livestock production.[47]

The spark that fueled such reaction has been commonly attributed to Rachel Carson's *Silent Spring* (1962), serialized in the *Denver Post*. Carson had documented the misuse of pesticides as well as misleading information that came from the chemical industry, its supporters in farm organizations, agricultural colleges, and the USDA. *Silent Spring* became an instant best-seller, suggesting that the values long proffered by conservationists and made more definitive by recent scientific discoveries had found fertile ground. *Silent Spring* became a clarion call for public action on a broad range of issues, which culminated in passage of the National Environmental Policy Act in 1969. Rachel Carson's exposé took place as the shift to a peacetime economy enabled more and more Americans to live, work, and play wherever they wanted. Colorado was a preferred destination.

A burgeoning, gentrifying population made its mark not only in and around the towns and cities along the Front Range but also in the scenic mountainous regions of the state. Once again, the federal government provided the impetus. In 1942 the US Army had sequestered forest service land and purchased private land in the Eagle River Valley as training ground for troops preparing to enter combat in the European Alps. Known as the Tenth Mountain Division, soldiers at Camp Hale famously learned to become expert ski mountaineers. At the same time, the US Navy had taken over the Hotel Colorado and its adjoining bathhouse in Glenwood Springs, converting that complex into a naval convalescent center. Bureau of Land Management (BLM) historian Steven F. Mehls had noted that the experiences of those soldiers and sailors and their subsequent word-of-mouth advertising did more than any paid promotional campaign to encourage tourism and associated development of Colorado's mountain towns. In addition, veterans taking advantage of the GI Bill to complete their education in Colorado's colleges and universities contributed immeasurably to the transformation of the state. From the viewpoint of the federal land agencies, city dwellers' increased recreational use on public lands helped garner public support of their respective budgets.[48]

Perhaps more than any other single individual, Walter Paepcke, president of Chicago-based Container Corporation of America, was the founder of the Colorado ski industry. Long attracted to Colorado, Paepcke discovered the abandoned mining town of Aspen. In 1947 his Aspen Skiing Company opened its first ski lift on public land. Two years later, as a leading patron of the arts and culture, he founded the Aspen Institute for Humanistic Studies. By 1960, Aspen had become recognized as among the finest tourist destinations in the world. The development of ski slopes on public land tempted longtime ranching families to sell their properties to housing and commercial developers.[49]

Along the Front Range, meanwhile, the decline in the number of farms and the ever-accelerating influx of new residents contributed to a major shift in extension service clientele. Chemical companies and implement dealerships had already supplanted extension expertise with respect to large farm and ranch operations. As a result, extension turned to urbanites, suburbanites, and exurbanites—providing advice on raising fruits, vegetables, ornamentals, trees, and shrubs. Extension specialist Herb Gundell became the leading public advocate for Colorado horticulture, popularizing the topic through his weekly column in the *Denver Post* and his weekly program on KOA radio. German by birth and brought up in Switzerland, Gundell had emigrated to the United States in 1939; three years later he enlisted in the army, trained at Camp Hale, and after the war used the GI Bill to earn degrees in horticulture at Colorado A&M.[50]

Between 1950 and 1970, the population of the four principal metropolitan Denver counties—Denver, Adams, Arapahoe, and Jefferson—nearly doubled, from 560,902 to 1,095,640. In neighboring counties, Boulder County's population increased by 270 percent, from 48,296 to 131,889; Larimer County's population doubled, from 43,554 to 89,900; and Weld County grew by nearly 50 percent, from 65,505 to 89,297. With the expanding population, a relatively new phenomenon appeared at the urban-rural fringe: commuters building homes on large lots or small acreages to enjoy "lifestyle" farming and keep a few horses. Some tracts were just over 35 acres, the minimum size to avoid the expenses of developing and maintaining a subdivision. Nonetheless, during this twenty-year period, Boulder, Larimer, and Weld Counties recorded almost 500 new rural subdivisions.[51]

Having had a start during the period of wartime mobilization, Jefferson County experienced the largest influx of newcomers. Between 1950 and 1970, county population increased by 450 percent, from 55,687 to 233,031. Land classified as agricultural declined by more than 300 percent, from 267,786 acres to 86,863 acres. In view of rapid residential and commercial development, a group of farsighted county residents once again persuaded county commissioners to call an election to consider a 0.5 of 1 percent sales tax for the purchase of parkland and open land. As a result of voter approval, the county created the Jefferson County Open Space Program, which between 1972 and 1981, acquired and conserved in perpetuity more than 11,000 acres in scenic, mostly forested areas of the county to provide recreational opportunities within hiking and biking distances of incorporated towns. On properties acquired with grazing rights, JeffCo Open Space allowed grazing to continue until contracts expired. To retain business support, JeffCo Open Space advocates made clear that their purpose was not to protect developable land from urban encroachment.[52]

In 1970, Governor John A. Love (R) managed to convince legislators to acknowledge the state's interest in local planning and enable the governor to appoint a statewide land-use commission, charged with classifying all lands in the state and determining how they might be used. With strong opposition from rural counties inspired by the farm bureau, the entire statewide planning effort was dropped. Farm bureau members had been led to believe that if land-use decisions were left to state officials, the next logical step would be national zoning dictated by the US Congress and the president. Failure of rural county commissioners to control second-home subdivisions and complaints by urban dwellers who wanted to conserve open land resulted in passage of the Local Government Land Use Control Enabling Act in 1974. Its companion, the Areas and Activities of State Interest Act, also passed in 1974, left the state with an advisory role only because rural legislators backed by the farm bureau had insisted that the state could not exercise veto power over local land-use decisions. The enabling act encouraged but did not require counties and municipalities to identify and designate certain geographic areas and activities of state interest. If a local government made such designation, then it had to set forth regulations, known as 1041 Regulations after the US House bill number, implementing a permitting process and issuing permits prior to development of such projects. Activities of state interest included site selection and development of new communities and, perhaps most consequential, efficient utilization of municipal and industrial water projects.[53]

Completion of the Eisenhower Tunnel in 1973, designed to carry four lanes of Interstate 70 traffic under the Continental Divide, frustrated further efforts to control development in and around mountain towns. Fishing and hunting on public lands, meanwhile, had deteriorated; ranchers limited trespass on their properties to those who could afford to pay; and public pressure for more recreation kept increasing. To restore and maintain once abundant wildlife, Governor McNichols recruited South Dakota state forester Harry R. Woodward Jr. as state game and fish director. Woodward was a pioneer in implementing the concept of multiple use while serving as manager of Custer State Park, the nation's largest state park, before multiple use became the slogan of federal land agencies. In Colorado, Woodward obtained game and fish commission approval to set the first fixed dates for hunting seasons and to allow year-round fishing, developed cooperative agreements with individual farmers and ranchers for the improvement of wildlife habitat on private lands, and assisted farmers whose cultivated fields were wintering grounds for big-game animals. Woodward considered habitat improvement an integral part of land and water conservation.[54]

The same urban, non-agricultural segment of the population that flocked to Colorado ski resorts, fishing spots, and hunting grounds was the driving force behind passage of landmark environmental legislation, beginning with the Wilderness Preservation Act of 1964 and continuing with the Clean Water Act of 1972 and beyond. Among agricultural organizations, the farm bureau was on record opposing any government regulations on land and water use, lobbying most forcefully against passage of the Clean Water Act—in particular, rules concerning non-point source pollution and runoff from chemically treated fields into rivers and streams. To refute a 1972 editorial in the *Denver Post* titled "Rural Lobby Can't Be Allowed to Wreck Environmental Bills," Colorado farm bureau president Lloyd Sommerville stated that "farmers and ranchers were environmentalists long before the word took on its present magical connotation."[55] Deliberately or otherwise, Sommerville misconstrued the intent of the environmental legislation that would serve as the future framework for public policy debate and legislation regarding the relationship between people and nature. By the 1960s, the term *environmentalist* had been adopted by urban-centered groups opposed to any development that might somehow damage air, land, and water resources. Speaking for the farm bureau, Sommerville described those groups as part of a deliberate effort to undermine agriculture's mission: providing a continual supply of food and fiber to the nation.

A more moderate view was expressed by state 4-H speech contest winner Robert Evans of DeBeque (Mesa County). To delegates at the 1965 CASCD, he spoke on the topic "Conservation, the Foundation for Your Future." Evans explained that 4-H and the CASCD shared the goal of "real conservation—using each acre of farm and ranch land within its capabilities and treating it according to its need for protection and improvement." This was a somewhat inelegant expression of the notion advanced by early progressives that all land should be put to its most productive use for the permanent good of all. In Evans's opinion, the ever-accelerating loss of farmland to residential, commercial, and infrastructure development made it all the more urgent to "check erosion, build up soil fertility, and regulate the supply of water so that the land will produce more and more, year after year." Soil conservation meant more than restoring and maintaining soil fertility; it required improving, even extending the soil's capacity to produce crops.[56]

The Reverend Daniel O. Parker, regional field secretary of the United Church of Christ, expanded on Robert Evans's notion of conservation, explaining to CACSD delegates that "soil stewardship in its completeness [must] take into account land being used with the greatest good for the greatest number."[57] Long gone were the days when landowners could do with their land whatever they wished; gone, too, were the days—as recent as the 1950s—when churches declared life on the family farm the only model for the good life, the family farm the sole sacred entity. Recognizing our growing dependency on each other, the limits of our land resources, and the melding of rural and urban living, he argued that one could no longer state that the old rural values are better or worse than urban values; both could be good or bad depending on how we use these values and how they affect both rural and urban people. "In light of the religious imperative that we have throughout all of scripture," Parker said, "there are certain basic issues which we have to face and face now." Old values remain valid in spirit but need to be reinterpreted to meet contemporary circumstances.[58]

6

The Specter of Nature's Limits

On September 10, 1976, Governor Richard D. Lamm (D) issued an executive order setting forth his administration's policy framework for managing growth and development in Colorado. Based on a well-considered conviction that resources in land, water, and living systems are finite, Lamm warned that policymakers could "no longer afford the luxury of ignoring the absolute, unavoidable, physical necessity of achieving balance between man and nature." In a practical sense, Lamm introduced into public life the ethic of stewardship expounded by naturalists ranging from George Perkins Marsh to Liberty Hyde Bailey to contemporaries Wendell Barry and Wes Jackson. Lamm's executive order, moreover, can be viewed as an early version of his thesis that the very future of life on earth depends on replacing our current "culture of growth" with an entirely novel "culture of limits." To be sure, the recognition of limited natural resources, especially water, is native to the arid West. But Lamm went beyond Captain Powell, Elwood Mead, Hugh Bennett, the New Dealers, and recent environmental prophets in emphasizing the political urgency of recognizing nature's limits.[1]

https://doi.org/10.5876/9781646422050.c006

Having served active duty as a first lieutenant at Fort Carson near Colorado Springs, Lamm, too, was attracted to Colorado by the magnificent countryside and spectacular skiing. After finishing law school at the University of California Berkeley, in 1961 he moved to Colorado where, in his opinion, it was still possible to prevent the haphazard, mindless development he had witnessed in California. No single sector dominating the Colorado economy, he thought, made it more likely to bring about systemic change. The immense and continuing popularity of the National Western Stock Show (est. 1906) at once symbolized nostalgia for Colorado's agricultural past and its yearning for a vibrant future economy without destroying what brought people to the state. Much taken by President John F. Kennedy's call to public service, Lamm had entered law practice with an eye on public service. His first foray was as volunteer lobbyist for the Colorado Federation of Garden Clubs against the billboard industry. In 1964, Lamm won a seat in the Colorado House of Representatives, representing neighborhoods surrounding the University of Denver. Having earned a reputation as an outspoken advocate for environmental protection, students at Colorado State University invited Lamm to keynote the first Earth Day "teach-in" celebration in April 1970. Earth Day was part of the movement that ushered in a decade of environmental legislation, most notably the National Environmental Protection Act, Clean Air Act, Water Quality Improvement Act, and Endangered Species Act. Lamm's embrace of that legislation did not ingratiate him with the farm bureau.[2]

Neither did his opposition to unfettered economic growth. In May 1970 the International Olympic Committee awarded the 1976 winter games to Denver. Arguing that the Olympics would promote business and thus lower taxes, the Colorado Olympic Commission convinced the state legislature to appropriate monies for the construction of infrastructure. Countering that the Olympics would be costly and would increase taxes and imperil Colorado's future, Lamm and his wife, Dottie, took out an advertisement in the *Denver Post* under the heading "Colorado, a Sellout," challenging the growth ethic of bigger is better: "Few of us want to live in another Los Angeles. Yet projections for the Front Range show a megalopolis extending from Fort Collins to Pueblo in a few short years." Inviting readers who agreed about the need for more public discussion of Colorado's future to contribute ten dollars or more, the Lamms were able to purchase several more advertisements. After failing to convince the Colorado General Assembly to withdraw funding for the Olympics, Lamm and two fellow legislators formed Citizens for Colorado's Future, attracting volunteers—environmentalists, conservationists, church and labor leaders—to collect the signatures necessary to put the issue on the ballot. In November 1972, voters approved the referendum against public funding by 59 percent; Denver withdrew its bid for the Olympics. Among agricultural organizations, only the farmers union had participated in the referendum coalition that, two years later, helped elect Lamm governor (1975–87).[3]

Lamm and his fellow urbanites may have held sentimental views about agriculture and rural living. But as a pragmatic idealist, Lamm considered farming and ranching indispensable to controlling growth. One year into his governorship, he told a farm bureau reporter that from a global food supply perspective, "the stork has out-flown the plow," which obliged everyone "to preserve every precious bit of agricultural land and water."[4] Toward that end, Lamm supported bipartisan legislation, ultimately found unconstitutional by the state, to prevent cities and towns from willy-nilly condemning agricultural water for municipal use. He opposed construction of C-470 around the southwestern edge of Denver, in part because it destroyed prime agricultural land, and opposed construction of Two Forks Dam in South Platte Canyon, which would have become the repository of half a million acre-feet of Western Slope agricultural water for Denver Water for municipal use.

When campaigning for governor, Lamm ran on a platform to ban new diversions of Western Slope waters as a means to control urban growth along the Front Range. To those who argued that after he arrived from California he simply wanted to build a wall around the state and pull up the drawbridge, Lamm countered by broadening the scope of debate. To members of the farm bureau, ever anxious about his views on environmental protection, he noted that, generally speaking, farmers and ranchers know they can graze only so many cattle in a field: "Likewise there are only so many people that can live comfortable, productive lives in Colorado. You can overgraze a field and you can also overgraze a state."[5]

To be sure, not all of Colorado was overgrazed. Governor Lamm advocated disbursing commerce and industry from the Front Range to counties that had been losing population. He knew about the French government's postwar initiatives to decentralize industry, thereby creating non-farm job opportunities in rural areas, and its belated recognition of the cultural

FIGURE 6.1. Governor Lamm with agricultural leaders, Denver, late 1970s. *Courtesy*, Agricultural and Natural Resources Archive, Colorado State University Libraries, Fort Collins.

and environmental value of conserving small-scale farms. Although an American governor could not dictate where private companies located their plants, Lamm did succeed in moving some state offices to Pueblo, Sterling, and other places beyond metro Denver. At the same time, he believed that keeping farms and ranches close to cities and mountain communities limited development and preserved open spaces, all of which depended on deliberate land-use planning.[6]

Despite modest attempts to control growth, economic development continued—seemingly unfettered by rules and regulations—throughout the tenure of the Lamm administration and that of his successor, Roy Romer (D, 1987–99), a former agricultural implement dealer in Prowers County and a strong advocate for farmers and ranchers. Even a cursory view of population and agricultural statistics from 1970 to 2000 illustrates the extent to which scientific discoveries and technical innovations had affected agriculture and rural communities. During that period, Colorado's rural population declined from 21.4 percent to 15.5 percent of the state's total population while the state's population nearly doubled, from 2.2 million to 4.3 million, with half of the population residing in the Denver metropolitan area. At roughly the same time, between 1969 and 1997, the market value of agricultural products statewide rose by 400 percent, to $4.5 billion annually; land classified as agricultural declined by 4.3 million acres, to 32.4 million acres. Statewide corn production increased by nearly 450 percent, to 131.5 million bushels; wheat production grew by roughly 200 percent, to 76.7 million bushels. Helping to fuel higher yields, expenditures on fertilizers and soil conditioners rose from $20.3 million to $124.3 million and petroleum products from $32.5 million to $121.6 million. Revenues from the agricultural sector as a percentage of gross state product continued to decline.[7]

Within that context, changing public attitudes reflected in federal legislation did alter the ways farming, ranching, and forestry were carried out. This change most immediately affected public lands, which comprised 35 percent of Colorado's land area. The Bureau of Land Management, the Forest Service, and other federal agencies could no longer manage their affairs with relative autonomy; consideration had to be given to the possible impact of human activity on the natural environment. The strongest advocates for federal legislation may have opposed any and all development on public lands, but the practical effect on agriculture was the development of several traditional approaches cloaked in new names—holistic range management, integrated range management, integrated pest management, ecosystem management, and others—with the common aim of managing a given landscape to restore and maintain its overall health. Among the earliest of these names was holistic range management, coined in the 1960s by Allan Savory to describe a method he had used to reverse desertification of grasslands in his native Rhodesia. As a result of his opposition to apartheid, he was forced to flee his native land, settling in Albuquerque, New Mexico, where he established the Allan Savory Center for Holistic Management to continue his conservation work. The Rocky Mountain Farmers Union, with backing from a large national foundation, tried but failed to adapt Savory's holistic approach to the revitalization of small, isolated rural communities in northeastern New Mexico.

Not without hope, sometime farmers union staffer Vencil Lee "Vince" Shively came up with a bold version of holistic management: revitalizing agriculture and communities in parts of the eight states overlaying the Ogallala aquifer, an area he referred to as the "groundwater commonwealth," reminiscent of John Wesley Powell's "watershed commonwealth." Shively had gotten his start on the family farm in Yuma County; he later became an ordained Congregational minister and earned a degree in planning from the University of Colorado. An eccentric and utopian of sorts, Shively had been taken by the communitarian ideas of the sociologist Amitai Etzioni. Following his untimely death, the farmers union incubated Ogallala Commons, Shively's idea of a nonprofit collaborative of agriculturists, local businesses, government agencies, and other interested parties. Ogallala Commons continues to take small steps to reverse the out-migration of people and money from the rural communities dependent on the Ogallala aquifer, ever cognizant of the prospect of entirely depleting that life-giving resource.[8]

Population growth and economic development on the Front Range had little immediate impact on agriculture and communities on the eastern plains, but that was not the case in the mountainous counties. Arguably the earliest, most imaginative effort to balance development and conservation, people and nature, took place in Routt County. It began in 1949 when John Fetcher, a Winnetka, Illinois, native and Harvard-trained electrical engineer employed by the Budd Company to build railroad cars in Europe, decided that he wanted to become a rancher. He and his brother Stanton purchased a 1,300-acre ranch in the Elk River Valley northwest of Steamboat Springs. Intrigued by the area's ski potential, Fetcher and three partners developed the first ski runs in the 1950s. By the late 1970s, their Steamboat Ski and Resort Corporation had spawned a real estate boom. Developers sought to acquire the very ranchland the Fetchers and their neighbors had come to enjoy.[9]

Among the neighbors was Stephen Stranahan, an Ohio industrialist who loved the West and owned the 500-acre Home Ranch. To protect that gem, he purchased the surrounding 4,000 acres. To protect the remaining 8,000 acres of Elk Valley floor, Stranahan approached the Fetchers and neighboring ranchers about limiting development by placing their properties in conservation easements. From a business standpoint, Stranahan understood that aesthetics, tranquility, and authenticity attracted high-end vacationers to his historic ranch, today one of the world's most luxurious dude ranches. Well-established land-rich and cash-poor ranch families felt apprehensive about their land-rich and cash-rich new neighbors; but they, too, wanted the valley to remain relatively free of development. Of particular concern was the threat of so-called ranchettes, 35-acre lots. In the Elk Valley, an acre valued for agriculture at $400 could bring 100 times that amount or more when sold for development—an attractive choice for a ranch family seeking to pay off debts, put children through college, and save for retirement. Between 1972 and 2000, Colorado lost 2 million acres of agricultural land to ranchettes. For affected counties, such developments posed serious financial burdens, generating insufficient tax revenues to cover required services such as road maintenance and fire protection. Moreover, especially in areas of high natural value, the introduction of noxious weeds, domesticated animals, and human-made obstructions to wildlife migration caused by development contributed to degradation of those values.[10]

Although they by no means generate the amount of income gained from selling land to developers, conservation easements are legally binding, permanently restrictive agreements meant to protect conservation and rural values. Landowners donate or sell development rights to a nonprofit land trust, which becomes responsible for the conservative management of the protected land; in return, landowners may receive a variety of tax benefits. The Fetchers donated their development rights to the American Farmland Trust; but recognizing working ranchers' distrust of outside conservation organizations, John Fetcher's son Jay became heavily engaged in the formation of the Colorado Cattlemen's Agricultural Land Trust—the nation's first land trust established by a livestock organization, the Colorado Cattlemen's Association. Meanwhile, Jay Fetcher and Steve Stranahan convinced some, but not all, valley ranchers to sign the "Upper Elk River Compact," which was no more than a voluntary, non-binding, one-page resolution in favor of "protective development." With the exception of a few previously identified low-impact building sites, signatories resolved to place their properties in conservation easements. Not entirely altruistic, the compact called for construction of affordable housing for ranch employees near the Clark general store with a post office at the center of the valley.[11]

Development is self-fulfilling to a large degree, and urbanites in search of healthy living exacerbated the pressure to develop more ranchland. Publicity about the Elk Valley compact prompted Routt County commissioners to embark on countywide land-use planning; realistically, if residents wanted to see open spaces preserved, they had to be willing to provide the monies to compete against developers. In 1996, Routt County residents first voted to tax

FIGURE 6.2. Conservation easement, Fetcher Ranch, Elk River Valley, ca. 1980. *Courtesy*, Jay Fetcher.

themselves so the county could purchase development rights—that is, make up to landowners the difference between the value of their land appraised as agricultural and its value when zoned for maximum allowable development. County voters decided to pay private landowners to maintain their lands in open spaces because residents prized those lands for their scenic, open space, and recreational values. Commissioners took planning one step further, adopting the concept of the land preservation subdivision, also known as cluster development. If a developer agreed to confine homes to a portion of a given property, meet certain design criteria, and keep the rest of the site in open space, the developer would be allowed to build one extra housing unit per 100 acres. Routt County and the Steamboat Springs municipality would take additional steps to slow the conversion of agricultural land to non-agricultural uses and to mitigate the adverse impacts of continuing development.[12]

Changes in ownership of large properties from established ranch families to wealthy non-agriculturists, as well as the influx of urbanites into rural subdivisions and Steamboat Springs, caused Routt County Extension Service to shift its focus. During his tenure as county agent (1989–2012), C. J.

Mucklow found himself in a strategic position to address the issues posed by development and demographic change. Extension staff turned much of their attention to assisting newcomers moving into rural subdivisions. "A Guide to Rural Living and Small-Scale Agriculture," compiled by Mucklow and his staff, was among the earliest of such ready-references to be adopted by extension staff in other Colorado counties facing similar situations.[13] Beyond the urgency of the issues, an increasingly urbane population sought out high-quality food, which contributed to a revival of interest in locally grown produce and livestock. Extension served as the incubator for the Community Agricultural Alliance, the creation of Marsha Daughenbaugh, a third-generation member of an Elk Valley ranch family that had decided that preserving the ranching heritage was worth more than selling out to developers. Under Daughenbaugh's leadership (1999–2018), the agricultural alliance connected forty-plus local producers with local restaurants, helped establish a year-round online food market so that even the smallest-scale producers could sell directly to consumers, and spawned the construction of commercial greenhouses that supplied fresh produce and a new USDA-certified processing facility for beef, lamb, and pork.[14]

In an unexpected sort of way, by placing a premium on open spaces and quality food, urbanites have contributed to the continuation of agriculture in Routt County. Partly by chance and partly by design, the extension service has managed to make use of nostalgic and sentimental interest in agriculture to address an equity gap, namely, limited access to high-quality food for some county residents. Outside Steamboat Springs but within the county where service industry employees live, between one-third and two-thirds of all children qualify for free or reduced school lunches. To be sure, with a short growing season, Routt County will likely never become food self-sufficient; further, the supermarkets in Steamboat Springs are unlikely to go out of business anytime soon. The story of agriculture in Routt County, replicated in and around other mountain towns and cities such as Boulder and Fort Collins, did indicate an appreciation of farming and ranching that went beyond maximum production to keep consumer prices low—a sense that there needed to be some reasonably harmonious connection between people and the land immediately surrounding them.[15]

But that understanding was far from universal. In 1988 National Farms, Inc. of Kansas City announced construction of Colorado's first, and the nation's

FIGURE 6.3. Farmers market, Steamboat Springs. *Courtesy*, Routt County Extension, Colorado State University, Fort Collins.

largest, hog confinement facility, designed to raise 300,000 hogs per year on the 25,500-acre 70 Ranch near Kersey, east of Greeley. Neighboring ranchers and local conservation groups opposed the facility's process for dissipating swine waste over the 70 Ranch, fearing pollution of groundwater and the nearby South Platte River, but to no avail. The arrival of National Farms, moreover, encouraged counties with declining populations and desperate for economic development to provide incentives to attract their own hog farms. Acknowledging that hog confinement facilities could not be prohibited outright and after failing to obtain regulatory legislation, farmers and ranchers, joined by conservationists and other interested parties, used the referendum approach. By a vote of nearly two to one, in November 1998, Colorado voters approved Amendment 14, which required confinement facilities to obtain permits, monitor their lagoons and land for pollutants, and provide assurance bonds. Two years later, citing regulations promulgated by Amendment 14, National Farms announced the closure of its 70 Ranch hog facility, leaving the state to clean up thousands of contaminated acres at a cost that far exceeded the value of the company's assurance bonds.[16]

Beyond the fields irrigated by trans-basin diversion, commodity growers on the eastern plains had continued to experience booms and busts brought on by global events as much as by changes in agricultural practices and the vagaries of weather. During the Nixon and Ford administrations (1969–77) farm prices rose, most notably following the 1972 agreement to sell an unprecedented 440 million bushels of wheat to the Soviet Union. Heartened by higher grain prices, growers invested heavily in new machinery and more land, often buying out their neighbors. The Organization of Arab Petroleum Exporting Countries embargo, beginning in 1973, caused fuel scarcities and rising fuel prices, which exacerbated farmer debt. Although the Carter administration (1977–81) appeared friendlier toward small producers than had the Nixon and Ford administrations, declining commodity prices were felt especially hard in winter wheat country where family farms still predominated.[17]

Out of despair, four Baca County wheat growers decided to take direct action to compel federal financial support at 100 percent of parity. More spontaneous than organized, naming themselves the American Agriculture Movement, their call for action quickly spread through twenty states. Baca County growers invited Secretary of Agriculture Robert S. Bergland, a former Minnesota wheat farmer and Washington lobbyist for the rural electric co-ops, to speak in Pueblo. On the scheduled day, September 22, 1977, about 2,000 farmers and ranchers drove their tractors and trucks to Pueblo and surrounded the auditorium where the secretary spoke. By then, the protestors had agreed among themselves to stop shipping their fall crops and to withhold fall planting unless the secretary guaranteed price supports that would allow them to earn fair profits. Clearly, the secretary could make no such promise. Baca County farmers then joined a multi-state "tractorcade" to Washington, directly petitioning the US Congress and the president. "A few trouble-makers gave us a bad name," recalled Two Buttes farmer Spike Ausmus, one of the "tractorcade" participants; the movement fizzled without obtaining any relief. In spite of its failure, the state of despair that brought on the movement serves to remind us about the ever-tenuous economic position of agriculturists who, perhaps more than any other profession, lack the means to control the prices of their products or the costs of production. The consolidation of grain elevators, flour mills, and suppliers meant even less control on the part of growers. Baca County's five grain elevators remained under the ownership of a farmers' cooperative headquartered in nearby Elkhart, Kansas; but elsewhere in Colorado, since the 1990s, Cargill Corporation has owned 80 percent of the grain elevators equipped to load unit trains. Elevator companies often specify the grains to be planted and provide growers with loans against future crops; the more uniformity in cultivars and traits (unique characteristics), the better for all parties concerned.[18]

Despite the immediate benefits of agricultural uniformity, some agriculturalists did recognize its long-term negative effects on soil health. During the Reagan years (1981–89), Congress in a bipartisan way took further steps to strengthen protection of land and water. Through the Food Security Act of 1985, Congress for the first time authorized appropriations for research, instruction, and extension activities in what became known as low-input sustainable agriculture. Sustainable practices went one step beyond the established policy of conservative use—producing the most food without impinging on future yields—to a more sophisticated notion of producing the most food without damaging the quality of the natural resources on which food production depends. The low-input program started with a very modest appropriation; that amount was succeeded in 1990 by the Sustainable Agriculture Research and Education Program (SARE), expanded to include training of extension personnel in sustainable practices and, most applicable to Colorado, conducting research on better integrating crop and livestock operations.[19]

Livestock is Colorado's most lucrative agricultural product, so most of the grain grown on the eastern plains is converted into livestock feed. As global demand for animal products continued to rise, scientists raised questions about the long-term environmental effects of applying larger and larger applications of anhydrous ammonia. That fertilizer had enabled farmers to plant the same monocultural crops, which produced greater and greater yields year after year. Smitten by the prospect of seemingly unlimited progress in technology, agriculturists and their supporters had forgotten that maintaining soil fertility over the long term required more than synthetic fertilizer. Indeed, synthetic fertilizers have been shown to disturb the natural cycle of soil nutrients replenished by the decomposition of organic plant and animal matter. That cycle remains indispensable for maintaining soil tilth, the soil's physical characteristics such as its capacity to hold moisture, and the proper balance of soil acidity and alkalinity. Into the 1990s, however, agriculturists remained confident that better cultivation practices, combined with gene modification to save on chemicals, would indefinitely ensure a sufficient supply of food for the nation and the world.[20] Continuing on the path pioneered by the Soil Conservation Service during the New Deal, the SARE program served as a means to attract attention to advanced techniques for improving soil quality, with the goal of higher profits for producers. Finding earth-friendly ways to increase dry-land crop yields was among the earliest and most long-standing SARE projects. At the southwestern Colorado research station near Cortez, experiment station scientists looked for new ways to save on water and minimize disturbing the soil through cultivation.[21]

Even before SARE, scientists at the Akron research station had been studying ways to increase grain yields under dry-land conditions. Best known to farmers on the eastern plains was soil scientist Bentley W. Greb, a Colorado State alumnus. But even he acknowledged the limits of technology, beginning one of his many articles by noting that the drought of 1977 had served as "a rude reminder that people do not yet control the whims of nature." Greb provided practical advice on improving the soil's ability to absorb and retain moisture, as well as on improving crops' use of water through fertilization, development of new varieties, and cultural manipulations. He was a "remarkable individual, wanting to tell you everything he knew," recalled Elbert County dry-land farmer Virgil Kochis, "but he couldn't easily explain to farmers what we needed to do and why."[22]

When William Pabor wrote in 1883 that once the "intelligent, systematic farmer" understood the proper methods for farming in semiarid Colorado, nature would take care of the rest, he was referring to irrigated, not dry-land, farming.[23] Ask Virgil Kochis. He will tell you most emphatically that dry-land farming is "stupid" because it is so utterly dependent on the weather. To say that average precipitation at Kochis Farms is 17 inches—of which 10 inches in rain—definitely misleads; there are good years, bad years, and very bad years. And yet, what began as a 160-acre diversified farm homesteaded in 1908 by Kochis's immigrant grandfather became a 10,000-acre grain-cattle operation that is still in the family's hands, now managed by Virgil and his son, Michael. With remarkable persistence over more than 100 years, the Kochis family has made adjustments in their operation to take advantage of marketing opportunities and, most fundamentally, in their recognition of the land's natural limits.

As a youth, Virgil Kochis recalled milking cows by hand, separating the cream for delivery to the creamery station in nearby Matheson and feeding the remainder to the hogs. With the advent of rural electrification, which made possible automatic milking machines and cold storage on the farm, his father increased the dairy herd from 12 to 85 head. At that point, milk was poured into 10-gallon containers and picked up by truckers for delivery to processors in Colorado Springs. By 1970, when Virgil and his brother Robert took over the farm, the dairy was no longer profitable; 4,500 acres in grass were turned over to beef cattle and 5,500 acres were planted in grains. The brothers adopted holistic grazing, moving cattle from pasture to pasture, and they only minimally tilled their grain fields. Abundant moisture during the 1980s encouraged them to replace wheat-fallow rotation with continuous cropping, a wheat-corn rotation, enough to build the cattle herd up to 200 head. For corn, Kochis had no choice but to purchase proprietary hybrid seed from commercial suppliers every year. Prior to spring planting, he "sterilized" the soil to eliminate weeds and potentially disease-carrying microorganisms. He contracted for application of custom-mixed fertilizer, roughly 40 percent nitrogen and 20 percent phosphorus depending on periodic soil tests. Minimal till and rotation helped prevent wind erosion. Depending on market forecasts, Kochis planted a feed grain other than corn, usually sorghum because of its superior drought tolerance. If he did leave a field truly fallow, he sprayed the stubble with herbicide after harvest in July and again in September, if needed.[24]

Once fully in charge of Kochis Farm operations and a wheat grower first and foremost, Kochis became a certified seed grower, a member of the Colorado Seed Growers Association. Taking a moment to consider the evolution of wheat growers coming together in association offers some insight into the increasingly sophisticated processes of production and marketing, leaving only partially answered the question of benefit to the land. Following passage of Capper-Volstead in 1922, which allowed farmers to associate for marketing purposes without violating anti-trust laws, wheat growers organized into the Colorado Association of Wheat Growers, initially to oppose exorbitant tariffs charged by the railroads. The Colorado Marketing Act of 1939 established the quasi-governmental framework that allowed producers of a particular commodity to join together for marketing, research, and education activities; but only in 1958 did wheat farmers vote in a referendum to assess themselves half a cent per bushel for those purposes. To collect and distribute assessments for carrying out the marketing order, wheat farmers nominated out of their own ranks and the commissioner of agriculture approved the selection of a governing board, the Colorado Wheat Administrative Committee. In 1963 the committee, with additional financial support from the milling industry, sponsored its first wheat research project at Colorado State University (CSU).[25]

Wheat research continued on a relatively modest level until the late 1980s, when the appearance in Colorado of the Russian wheat aphid (*Diuraphis noxia*) threatened severe crop losses. In response to that menace, the administrative committee and the wheat growers association successfully sought special state appropriations for CSU researchers to find ways to safely exterminate the aphid and develop an aphid-resistant wheat variety. Through a second referendum in 1988, wheat growers increased their assessment from half a cent to 1 cent per bushel. Within six years wheat breeders had developed a new cultivar, appropriately named 'Halt,' that was resistant to the wheat aphid. As was standard for a new release, agronomists at the experiment station supervised the production of seed stock for distribution to certified seed growers.[26]

With the introduction of 'Halt,' Colorado wheat farmers and CSU researchers entered into a unique relationship: wheat farmers made up the governing board of a new nonprofit organization, the Colorado Wheat Research Foundation (CWRF), established for the purpose of taking ownership of newly developed wheat varieties. The CWRF took exclusive responsibility for protecting the variety's intellectual property rights, collecting royalties from sales of the variety, and reinvesting those royalties in support of further wheat and wheat-related research. In effect, this positioned the CWRF to support the development, propagation, production, and distribution of all wheat varieties and traits developed at CSU. In 2012 the CWRF established the patented PlainsGold brand to market its varieties beyond Colorado. Further blurring the demarcation of public and private, in 2001 CSU, through the CWRF, became the first public university to release a wheat variety that included a patented trait (Clearfield) developed by private industry, in this case by BASF, a German chemical company. Beginning in 2008, again through the CWRF, CSU researchers developed an herbicide-tolerant trait to control winter annual grassy weeds such as cheatgrass (*Bromus tectorum* L.), which germinate in fall and winter and grow actively in spring. In 2018 the CWRF engaged Albaugh Global, which had patented the synthetic herbicide linked to the CSU-invented trait, and partnered with Groupe Limagrain, an agricultural cooperative, which became the exclusive marketer worldwide for both the trait and the herbicide. In addition, the venerable Vilmorin SA, Groupe Limagrain's seed division, obtained worldwide rights to transfer the trait to other varieties and license the trait-plus-herbicide system to other companies.[27]

The use of the CoAxium Wheat Production System, as the trait-plus-herbicide would be known, combined with minimal tillage should allow for fewer sprayer passes, reductions in fuel consumption, and buildup and retention of organic matter—all leading to improved soil health. It remains to be seen, however, whether this systems approach to wheat cultivation simply extends the traditional collaboration between land-grant universities and industrialized agriculture to increase profits and lower production costs or represents something more imaginative meant to improve soil health and the ecosystem generally. Nevertheless, the development and marketing of CoAxium marks a significant step by CSU in the world of agricultural science and technology, reinforcing the notion of the global interconnectedness of all aspects of agriculture.[28]

Even before the development of CoAxium, Virgil Kochis estimates that wheat yields over the past half-century have increased from 20 bushels to 60 bushels per acre. He attributes those increases to minimum tillage, keeping harvest residue on the soil surface, and catchment of more snowmelt. On his farm, soil balance between acidity and alkalinity is back to pH 6.5, considered optimal for winter wheat; organic matter content measures 1.5 percent of soil composition, natural for the eastern plains; and synthetic fertilizer is no longer required. For the general visitor to a dry-land wheat farm, it is close to miraculous that with three monstrous machines—a sprayer, planter, and harvester—one person can singlehandedly farm many thousands of acres.[29]

Kochis wanted to continue the family tradition of dry-land farming no matter the risks, so he welcomed the opportunity to diversify his income by leasing land to Xcel Energy, the state's principal public utility, for wind development. In an effort to reach its goal of 100 percent carbon-free electricity by 2050, Xcel planned to generate 600 megawatts of electricity by building and operating 300 wind turbines and transmission lines on two separate wind farms across 95,000 acres in Cheyenne, Elbert, Kit Carson, and Lincoln Counties. The Rush Creek Wind Project became the largest wind project in

FIGURE 6.4. Raising cattle, generating energy, 2019. *Courtesy*, Bob Kjelland, Rocky Mountain Farmers Union.

Colorado, sufficient to power 325,000 homes and eliminate 1 million tons of carbon pollution annually.[30]

To be sure, the use of wind to generate electricity on farms and ranches goes back generations, even before the New Deal. But only in 2002 did land agents representing energy companies begin to approach farm and ranch families on Colorado's eastern plans about leasing land for turbines. Proposed contracts varied in terms of duration, easements, risks, and payments—about which agriculturists were ill-prepared to negotiate. By the time of the Rush Creek project, landowners were aware of safeguards and pitfalls. Virgil Kochis asked a well-known Denver attorney specializing in renewable energy development to negotiate a lease agreement so the company could determine the feasibility of a wind farm. Once the wind farm was built, the company's agreement with participating landowners included annual rental of $5 per acre for land without turbines and a royalty payment on production for land with turbines. By 2018 thirty turbines, each on a separate acre, were operating on the Kochis farm. The family sold 40 acres for Xcel's operations and maintenance facility and a transmission substation. As to the risks of a wind farm, Jan Kochis told a local reporter that adverse weather conditions did not matter: "As long as the wind blows, we're producing energy and earning a percent[age] of the revenues generated. It is our new cash crop."[31]

The Kochis family and their neighbors were fortunate in their proximity to electric customers along the Front Range and in their dealings with a company that had the capital to build the necessary transmission lines. Less favored were farmers and ranchers in Baca County, perhaps the windiest county in Colorado. There, local landowners tried to build and operate their own wind farms. In the opinion of County Commissioner Spike Ausmus, the ability to attract a major wind project depended on the US Forest Service granting a right-of-way for transmission lines to cross the Comanche National Grasslands, the most direct path to populated areas in southern Colorado. But the obstacles to developing wind power in Baca and other southeastern counties were minor compared to the reality of declining water tables and diminished pumping rates. Without state regulations, Ausmus continued, the only limit on groundwater consumption would occur when pumping costs exceeded irrigation benefits.[32]

To make economical use of expensive water, some Baca County farmers have begun experimenting with high-value crops such as onions, cantaloupe, squashes, and other vegetables. The Baca County Conservation District and the Natural Resources Conservation Service (formerly the SCS) offered to pay up to 75 percent (within certain limits) of the cost to install sub-surface drip irrigation systems. Basically, these systems consist of buried plastic tubes with regularly spaced emitters sending water drop by drop directly into root zones; timing is regulated by an above-surface control mechanism. Sub-surface drip irrigation is costly to install and requires cultivation with specially designed disks to maintain planting beds in their same locations and to prevent damage to sub-surface drip lines. But it does offer the great advantage of eliminating surface runoff, reducing evaporation, and conserving soil nutrients. Compared to furrow or flood irrigation that is 50 percent efficient in its use of water, sub-surface drip irrigation is 98 percent efficient.[33]

Although common in the Middle East and elsewhere, sub-surface drip irrigation was new to Colorado. As part of the SARE program, CSU irrigation specialist James Valliant recruited two young Baca County farmers, Brent and Penni Morris, to explore ways of making sub-surface irrigation economically viable. The Morrises operated their farm on 1,600 dry-land acres and

1,780 irrigated acres. Seeking to cut costs, they agreed to try growing onions on 170 acres using sub-surface drip irrigation. To prevent seeds shallowly sown from blowing away and to protect seedlings from wind damage, the Morrises planted small grains in the middle of growing beds and furrows, chemically terminating the grains once the onions were established. During the growing season, they experimented with a combination of the mineral zeolite and the synthetic hydrogel to help spread and retain water, with no apparent improvement in soil structure or onion yields.[34]

Residents of the eastern plains have acknowledged at least tacitly that no amount of technology can reduce the need for irrigation to the point where groundwater pumped out balances the aquifer's natural rate of recharge, much less allows water tables to increase. To date, no one has devised a viable plan for what will happen when the aquifer dries up. New Dealers had experimented unsuccessfully with moving farmers off sub-marginal lands and providing financial support for their resettlement, restoring those lands with native grasses and building a non-agricultural economy. In the late 1980s Deborah E. Popper, a graduate student at Rutgers University in New Jersey, caused an uproar after publishing an essay suggesting that the federal government buy up the entire Great Plains, convert them into a vast park-like "buffalo commons," and then create a federally owned corporation similar to the Tennessee Valley Authority to fund region-wide community and economic redevelopment. On a more realistic track for adapting agriculture to the land, Kansas native and geneticist Wes Jackson brought together a group of imaginative scientists to form a nonprofit research organization, the Land Institute. His dream: to develop "Natural Systems Agriculture" in which perennial grains would produce high yields of seeds to replace wheat and other grass annuals. To date, institute scientists have domesticated an intermediate wheatgrass (*Thinopyrum intermedium* [Host] Barkworth & D. R. Dewey) into a perennial grain, calling it Kernza®; the seed is distributed to researchers for growing under various conditions in Minnesota, New York, Colorado, and elsewhere. Patagonia Provisions was the first to develop a commercial product made from Kernza®.[35]

In contrast to dry-land farmers, agriculturists with land connected to natural surface streams (tributary waters) faced different challenges. Recall that under the Colorado system, water diverted for irrigation cannot interfere with priority rights and must be used for beneficial purposes. Upstream irrigators are required to return "unused" water back into the stream, allowing downstream irrigators to use their allotments. Those who praise the conservative value of drip irrigation and complain about the apparent water waste in flooded furrows should take note that about half of flood irrigation waters gradually filter down deep into the soil, returning to the stream bed. By law, drip irrigators must make up for water not "lost" to percolation, whether through water exchanges, financial payments, or other means facilitated by their local water conservancy districts; superficially, this constitutes a disincentive to conservation.

A far more insidious disincentive is Front Range cities using a variety of approaches, some direct and some clandestine, to secure water for their growing populations. This undermines land and water conservation activities initiated by farmers and ranchers, in effect, threatening the very existence of agriculture on some of Colorado's most productive land. During the 1960s, Aurora began purchasing water rights from ranchers in South Park and farmers in the Arkansas Valley. Aurora's developers, most notably the Tower family, sought to prolong the World War II economic boom and outcompete Denver for business and industry.[36]

Arguably the most aggressive municipality in the search for more water was Thornton, a rapidly growing suburban community just north of Denver. In a case that would have far-reaching consequences for Colorado agriculture, the City of Thornton initiated condemnation proceedings in 1973 against the Farmers Reservoir and Irrigation Company. The city had offered $3,920 per share for a total of $9.3 million for the purchase of water and water rights and ditches and ditch rights used to irrigate 20,000 acres, mostly in Adams County. Thornton already owned 150 shares appraised by independent water consultants at $2,600 per share, but farmer shareholders estimated that the water was worth at least twice that amount. Only 3 farmer shareholders out of 271 agreed to Thornton's offer, which allowed the city to proceed with condemnation. Attorneys for Thornton appeared in district court, arguing that the city could invoke the right of eminent domain. While acknowledging the constitutionally mandated priority given to domestic over agricultural use, attorneys for the farmers argued that most of the water sought by the city would be used for watering lawns. If watering lawns was defined as "domestic use," then keeping lawns green would have a higher priority than growing crops. The court agreed with the farmers.[37]

While *City of Thornton v. Farmers Reservoir and Irrigation Company* was still in litigation, the legislature passed, and newly elected governor Richard Lamm signed, the Colorado Water Rights Condemnation Act, making it more difficult for a municipality to use condemnation as a means to obtain water. The act required a municipality to prepare a detailed community growth plan accompanied by a statement on the overall impact of condemnation and further restricted a municipality by limiting condemnation to a period not to exceed fifteen years.

Because of the case's broad implications, the state supreme court agreed to consider Thornton's appeal. By a vote of five to two, the high court reversed the district court's decision. The court majority cited the state constitution (Art. 20, Sec. 1), which gave a home-rule municipality "the power within or without its territorial limits to construct, condemn and purchase, acquire, lease, add to, maintain, conduct and operate water works . . . as in taking land for public use by right of eminent domain." Thus the legislature had no power to enact any law that denied a right specifically granted by the state constitution. The Water Rights Condemnation Act of 1975 was declared unconstitutional. In the minority's opinion, the state constitution granted home-rule municipalities only the power to control matters of local and municipal concern. Thornton's proposal was a matter of statewide concern and therefore subject to statewide regulation. The minority justices wrote: "Expanding water requirements of growing municipal populations are causing much of Colorado's scarce water, previously devoted to agricultural use, to be acquired for municipal use. Municipal condemnation of water rights necessarily has a tremendous impact upon the economies of agricultural areas and disrupts farm operations which are far distant from the condemning city." Governor Lamm could not have stated better the urgency for statewide land-use and controlled growth.[38]

Meanwhile, the continuing population growth in the metro Denver counties placed additional strains on existing water supplies. With nearly half of all domestic water consumption used for irrigating lawns, the Denver Board of Water Commissioners found it expedient to initiate a collaborative effort with the Associated Landscape Contractors of Colorado to encourage water conservation. A key participant in this effort was James R. Feucht, an extension professor of horticulture stationed in Denver, a member of the organizing committee, and the principal missionary for the agreed-upon approach to be known as xeriscaping—derived from the classical Greek "xeric," meaning arid conditions. Trademarked by Denver Water (as the board became known), the committee defined xeriscaping as "water conservation through creative landscaping," which included using drought-resistant grasses, low-water–requiring plants, soil amendments such as compost, mulches to reduce evaporation and keep soil cool, and efficient irrigation such as drip irrigation systems. Since the 1980s, xeriscaping has been a boon to the green industry and has interested more and more residents in landscaping with plants adapted to semiarid conditions.[39]

Front Range municipalities, nonetheless, continued their search for more water. During the 1980s, Thornton had initiated condemnation proceedings against two more irrigation companies, both in Adams County, and had taken other, less direct approaches to obtaining water as well. During that decade, Thornton purchased 21,000 acres of farmland between Fort Collins and Ault and after years of litigation secured those water rights to the Cache le Poudre River that went with the farmland, all of which enabled Thornton to double its population. But perhaps the most impressive evidence of transferring water from agricultural to municipal use was seen in Crowley County, where 50,000 irrigated acres had once supported 500–600 farm families; by the year 2000, the numbers had dwindled to fewer than 5,000 irrigated acres supporting 12 families, with the remaining acres allowed to dry up.[40] "When water leaves, the community dies," observed Travis Walker, the pastor of several Arkansas Valley Methodist congregations. He explained that towns may not disappear physically, although that has happened, but they lose their sense of cohesiveness. Pastor Walker heard from more than one farmer, "I'd rather have a neighbor than my neighbor's land."[41]

To put the "buy and dry" movement into agricultural context, one needs to go back to the late 1880s when a land speculator named T. C. Henry envisioned a network of ditches and laterals to irrigate 1 million acres in the lower Arkansas Valley from a headgate on the Arkansas River near Boone (Pueblo County) through Crowley and Prowers Counties to the Kansas state line. Like so many early development projects in the West, the Colorado Canal missed its goal, covering just 50,000 acres in Crowley County. As farmers sought to grow high-yielding crops and extend their growing season, they looked to the Western Slope for supplemental water. In the 1920s shareholders in the Colorado Canal Company paid for the enlargement of two natural lakes near Leadville, creating Twin Lakes Reservoir, which allowed for storage of trans-basin water from the Roaring Fork of the Colorado River. The reservoir was further enlarged as part of the controversial Fryingpan-Arkansas Project, authorized in 1962 and completed in 1981; the project was controversial because its stated purpose was to divert, store, and deliver water primarily for municipal and industrial use.[42]

No one was a more outspoken critic of diverting water to cities than Crowley County farmer Orville Tomky. In the 1950s he had bought a 160-acre farm near Olney Springs, planted to sugar beets, corn, alfalfa, tomatoes, and melons; at the time, it was enough land to support his family. As an irrigator and a shareholder, Tomky served as a director for the Colorado Canal Company and the Twin Lakes Reservoir and Canal Company. Starting in the late 1960s, as farm prices declined and farm debts rose, selling water rights became an attractive way for a farmer to get out of agriculture, and retire. To keep water in agriculture, Tomky offered to buy shares from his fellow farmers but could not begin to match the prices offered by companies with benign names such as Crowley County Land and Development Company, whose principals remain unknown to this day. Essentially shell companies, they kept land in agriculture for one or two years, then sold land and water rights to the cities.[43] Most disheartening to Tomky were the farmers who put personal gain above the common good. "The ones who sold their water, sold out their county," he told a reporter for the *Chicago Tribune*. "It's a moral issue," he added. "Do you have a right to sell out your neighbors? Your town?" He considered selling water to the cities to be betraying fellow farmers and undermining the system of mutual aid. Traditionally, the canal companies had served to ensure the classically liberal balance between individual rights and community responsibilities, allowing farmers to do what they wished with their operations as long as they did not impinge on the rights of their neighbors. Since shareholders in mutual aid companies vote according to the number of shares each owns, once the cities gained the vast majority of shares, the system of mutual aid vanished.[44]

From an amoral, strictly business perspective, farmers who sold their water rights acted as willing sellers, recognizing that they earned more from selling water rights than from selling crops for which they retained the water. But that is precisely how the market for water rights was designed, moving water rights to what would become known as their highest and best use. Nonetheless, as part of purchase agreements, the cities were obligated to put dried-up farmland into grasses. Until recently, those agreements did not specify how and by whom the re-vegetated lands were to be cared for over the long term; in addition, earlier agreements did not require the state to pay counties in lieu of tax revenues gained from assessments on productive farms. Although dry-land farming seemed like the perfect answer for restoring the land to agricultural uses, it has not succeeded, explains Matthew D. Heimerich, Orville Tomky's son-in-law. After nearly a century and a half of irrigated farming, soils have changed: "It's not that healthy, native soil that you would see on the prairie." Instead, soils have become silty, greatly reducing their ability to sustain even dry-land cultivation.[45]

Today's traveler through Crowley County will see thousands of acres in weeds and other noxious plants, exacerbating the challenges of cultivation for the few remaining irrigators, among them the Tomkys. A New Yorker by birth, Heimerich was first attracted to the West by the mountains and ski slopes. He met his future wife, Karen, a nurse practitioner now with her own clinic in Ordway. Karen and Matt, a history major in college with no experience in agriculture, bought a small house and 100 acres of irrigated farmland; Orville Tomky contributed water and equipment and, most important, served as Heimerich's mentor on agriculture. Since that purchase, the Heimerichs have expanded to 320 acres, combining their farmstead with those of Karen's two brothers and two nephews into a single family farm corporation. In the spirit of his father-in-law's commitment to agriculture and rural life, Heimerich served three terms as Crowley County commissioner and recently joined the Palmer Land Trust (today, Palmer Land Conservancy) as conservation director for the lower Arkansas Valley.[46]

Absent legislatively mandated regional land-use planning, the Palmer Trust took the initiative to bring together representatives of valley cities, towns, and counties, which resulted in the composition and implementation of a comprehensive conservation plan for the lower Arkansas Valley—an area of approximately 1.75 million acres stretching from Cañon City (Fremont County) on the west to Rocky Ford (Otero County) on the east.[47] Started in 1977 by a small group of Colorado Springs residents seeking to preserve open spaces for aesthetic reasons and recreation, the Palmer Trust expanded its mission to a more all-encompassing approach, seeking ways to restore and preserve the land and at the same time ensure economic prosperity and community livability. A case in point: the City of Pueblo, a major purchaser of water rights and farmland. With Pueblo expressing some willingness to lease as yet unneeded water back to farmers, the Palmer Trust has sought to convince water authorities to allow temporary transfer of that water onto the most productive farmlands without going through the complexities of the water courts. Pueblo has a special interest in conserving agriculture in the Arkansas Valley. Within a 50-mile radius, farms both large and small provide residents with an amount of nutritious and affordable food unmatched by other cities in the arid West. The matter of food security becomes of increasing concern as industrial agriculture focuses on just a few varieties of commodity crops and distributors of vegetables and fruits depend more and more on imported produce.[48]

Recall Major John Wesley Powell's thesis that by supplying irrigable water to the arid West, the federal government would ensure that the region would develop as and remain a place of small farms and small towns and its implication that the arid West could never support an urban civilization. Thus Matt Heimerich and his colleagues represent a contemporary cadre of pragmatic idealists committed to the proposition that by learning how to live within nature's limits, an urban civilization can indeed thrive. The juncture of rural agrarian and urban industrial posits several ironies, encapsulated in the story of the Arkansas Valley farm family that sends its children to college in the city that has preempted the source of the family's livelihood or the residents of suburban, ex-urban, or rural housing developments located on lands that once produced the locally grown fresh fruits, vegetables, and meats for which they anxiously yearn.[49]

Robert F. Crifasi, research scientist and former water resources administrator for the City of Boulder, has directed attention to a rarely discussed aspect of water development that helps explain why Captain Powell's prediction has yet to materialize: "Because much of our food is grown far from Colorado cities, we in effect use and import water resources from far away." Unmeasured and hidden from public view, "this remote water is bound up in food and other commodities produced elsewhere." Considering that 85 percent of Colorado water still goes to agriculture, Crifasi noted that applying the principle of highest and best use would allow Colorado's population to continuing to grow into the distant future. Following the demise of the Two Forks Dam, conservationists wanted to believe that population growth and development on the Front Range would slow down. That did not occur, leading Crifasi to make the point that the critical issue was not a lack of water but reallocation of water, which takes us back to Governor Lamm's dream of achieving a balance between people and nature: controlling development while conserving those values that have made Colorado such an attractive place to live and work.[50]

Although such a balance remains elusive, no county has done more to reach that goal than Boulder County. Similar to Golden, located in neighboring Jefferson County where Clear Creek flows out of the foothills, Boulder is located at the spot where Boulder Creek and the nearby South St. Vrain River flow out of the hills, at one time also irrigating prime agricultural lands. Between 1959 and 1974, Boulder County led the state in the number of acres converted from agricultural to non-agricultural use. In 1978 the county adopted a land-use plan focused on preserving agricultural land to slow down and control development. The non-urban planned unit development allowed the owner of agricultural land to place two houses rather than one on 35 acres, provided that 75 percent of that acreage was placed in a conservation easement held and managed by the county. After recognizing that the approach unexpectedly facilitated the development of large homes on big lots, the county increased minimum requirements from 35 acres to 320 acres. In addition, the county approved the transfer of development rights from land of high agricultural or environmental values to land, not necessarily contiguous, where development would be less harmful to those values; it also approved the outright purchase of development rights, conservation easements to be held by the county. In all cases, water rights had to remain with the protected land and be deeded to the county. Boulder County residents voted to tax themselves for the specific purpose of protecting agricultural land and open spaces.[51]

Further away from the foothills, mostly east of Interstate 25, development appeared to take priority over conserving open land, causing municipalities to continue to search for more water. Starting in the 1970s, the farmers and ranchers who ran large commercial operations increasingly experienced uncertainty about their ability to resist urban encroachment. Making tenuous situations worse, agriculturists especially around Denver and in northern Colorado found themselves the subjects of lawsuits filed by ex-urban and suburban neighbors who objected to farm noises and odors, machinery and livestock on public roads, herbicides and pesticides, and other normal agricultural practices. "Right to Farm" legislation enacted during the 1980s

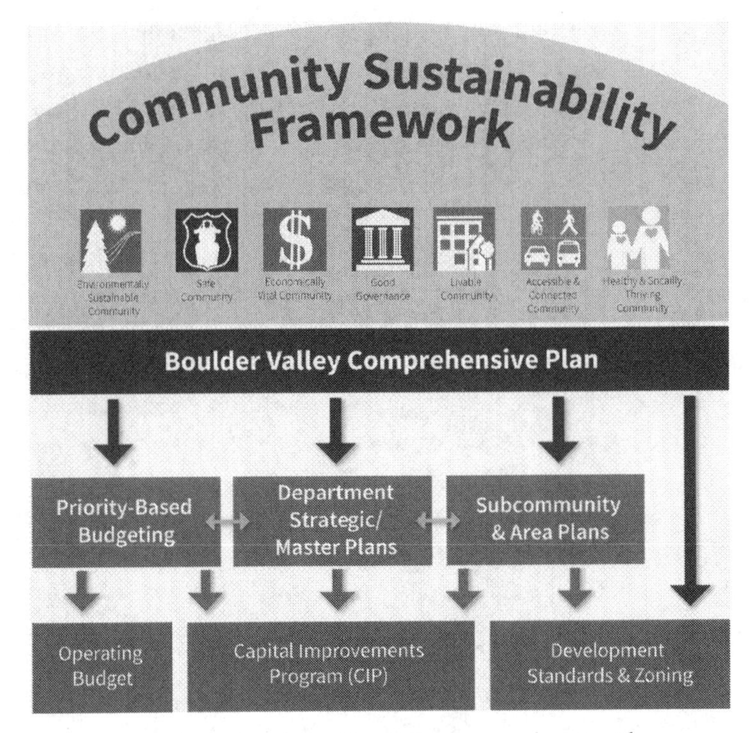

FIGURE 6.5. Schematic of Boulder Valley Comprehensive Plan, 2010. *Courtesy*, City of Boulder, Colorado, Comprehensive Planning.

limited the circumstances under which a farm or ranch could be deemed a nuisance and affirmed that local governments could adopt ordinances to provide additional protection to agriculturists.[52]

Meanwhile, the hodgepodge of residential development, publicly owned open space, and farms and ranches on city perimeters did create opportunities for smaller-scale, more intensive high-value farming—cultivating the fresh produce and raising the livestock increasingly in demand by those urban dwellers who were seeking locally grown, healthful food. In addition, the growing disparity between those urban dwellers who had access to wholesome food and those who did not caused Denver city officials and nonprofit leaders to begin experiments in urban agriculture—adapting the Victory Garden approach on school properties and rooftops and in vacant lots, backyards, and other small areas—as a way to address the problem of malnutrition, overcome the scarcity of grocers selling affordable and high-quality fresh foods, and contribute to neighborhood revitalization. Because of the generally small-scale nature of farming within and on the periphery of urban areas, cultivating organically, without artificial soil amendments, became a practical technique by the 1980s, although official criteria for defining products as organic did not occur until a decade later.

Like many movements once considered eccentric, the rehabilitation of organic farming as a modern method of cultivation began in England; it was organic in the general sense of connecting all parts into a whole. Its underlying theory was traceable to Charles Darwin's *The Formation of Vegetable Mould through the Action of Worms* (1881): all agriculture depends on humus, the organic component of the soil—organic in the specific chemical sense of containing carbon formed by the decomposition of plant materials. The English botanist Sir Albert Howard (1873–1947) is considered the founder of modern organic farming. He spent most of his career as an agricultural adviser in British India. There he observed how farmers created a closed "virtuous circle": sowing seeds, cultivating crops, saving some seeds for planting, returning crop residue to the soil where microorganisms converted plant materials into humus, and continuing the cycle indefinitely. Sir Albert undoubtedly envisioned a parallel between the virtuous circle of agriculture and the philosophical wheel of life that underlies Buddhist philosophy. At the risk of oversimplification, organic agriculture, based on a cyclical principle, operates within the limits of nature; conventional agriculture, based on a linear principle, depends on perpetual growth, ever seeking to override nature.[53]

Sir Albert's seminal *Agricultural Testament* (1940) appeared just as cultivators of commodity crops began to depend on synthetics, all part of the wartime effort to greatly increase yields. More than a handbook for organic farming, his testament served as a call to action for farmers to stick with natural fertilizers, herbicides, and pesticides. Sir Albert's most outspoken disciple in the United States was an eccentric entrepreneur, Jerome Irving Cohen (1898–1971), a grocer's son who grew up in an immigrant neighborhood on the lower east side of Manhattan. After engaging in several business ventures, Cohen, who anglicized his name to Rodale, bought a farm in Emmaus, Pennsylvania, where he tried to put Sir Albert's cultivation method into practice. He also created a publishing and natural health products business that appealed to consumers seeking healthier foods. In an unsentimental study of Rodale and his empire, historian Andrew N. Case has noted that "both the

Rodale Press and marketplace environmentalism succeeded in altering the aisles of supermarkets, but they could barely make a dent out in the fields."[54]

Taking advantage of the farm crisis, in the late 1970s Rodale inaugurated *New Farm*, a monthly publication based on the premise that organic farming reduced energy, fertilizer, and other costs while increasing farmer profits. With a few isolated exceptions, the Rodale approach remained virtually unknown among farmers in Colorado. But the Rodale message about health and "natural" foods did contribute to increasing consumer interest in organic products. In 1979, Robert S. Bergland became the first secretary of agriculture to turn serious attention toward organic farmers. He commissioned a study of organic farming practices, based in part on a Rodale survey of *New Farm* readers. The ensuing report documented growing concern about the use and overuse of agricultural chemicals and their adverse effect on soil health: "Consequently, many feel that a shift to some degree from conventional toward organic farming would alleviate some of these adverse effects, and in the long term would ensure a more stable, sustainable, and profitable agricultural system." Although some organic farming practices had been promoted by the US Department of Agriculture (USDA) since the 1930s, the report explained that modern organic farmers had not simply reverted to ancient practices. Contrary to popular belief, the vast majority were taking full advantage of agricultural technology—to this day, a point often lost on critics of organic farming.[55] Recommendations made in the Bergland report were officially adopted in 1985 through the Low-Input Sustainable Agriculture Program, succeeded by the Sustainable Agriculture Research and Education (SARE) Program.

In Colorado, Saguache County rancher Mel Coleman Sr. and Larimer County farmers Lewis Grant and his son Andrew led the effort to organize the Colorado Organic Growers Association, which in 1989 lobbied the legislature to authorize and fund a state-administered organic certification program. Legislative leaders assigned the bill in question to the House Military Affairs Committee, hoping that would guarantee its demise; but the bill passed and was sent to the full house after the farm bureau lobbyist expressed no opposition. The Colorado Department of Agriculture became the first state agency in the nation to begin such a program. Other states followed Colorado, each adopting certification standards tailored to its own organic practices and climatic conditions. Growers soon recognized that because

FIGURE 6.6. Official USDA certified organic product label (seal). https://www.ams.usda.gov/rules-regulations/organic/organic-seal.

of the nature of interstate commerce and so as not to confuse consumers, there needed to be a national certification program with a common definition of organic, with the ability to revise that definition as new technologies came into practice and to provide certification oversight and enforcement. In 1990 President George H.W. Bush signed the Organic Foods Production Act, which led to the National Organic Program (NOP). The process for implementing rules and regulations was completed in 2000; two years later, the Colorado Department of Agriculture became a USDA-approved certifier to the NOP standards. While not all organic farmers went through the certification process, the organic label did provide consumers with verification of cultivation methods and gave producers a reason to charge premium prices.[56]

That was of special significance to small-scale farmers, something that attracted the attention of the Rocky Mountain Farmers Union. Its principal advocate for organic farming was Longmont native David E. Carter, a veteran publicist and part of the organization's leadership at both the national and state levels. Through extensive lobbying of state legislators and the congressional delegation during the Lamm administration, he had become well-known to directors and staff of the National Farmers Union. During his tenure as national secretary (1987–93), when he worked with state divisions, Carter met farmers seeking ways to reduce reliance on synthetic additives, in particular New England dairy farmers looking for profitable ways to abandon the use of antibiotics in improving milk productivity. He helped write the rules to carry out the national organic standards. From his national perspective, Carter came to realize that ground zero (his term) of the natural and organic foods industry was Boulder, Colorado. There, he became acquainted with the pioneers in the natural and organic foods distribution business, in particular Mel Coleman Sr., founder of Coleman Natural Meats. Carter envisioned organic agriculture as a way to save small to mid-scale farms.[57]

In November 1993, Dave Carter was elected president of the Rocky Mountain Farmers Union; during his seven-year tenure, he confirmed by his actions a renewed commitment to the organization's founding principles of economic and social justice. He understood that the farmers union was more than a farm organization and even more than a farmers' organization, an insight that positioned his organization to take on statewide leadership in the incipient movement for food security through the development of cooperatives. In early 1994, while listening to a small group of San Luis Valley farmers and ranchers complain about persistently low livestock prices and a market in which they could not compete, Carter sensed an opportunity for organic farming that he believed could benefit from the cooperative structure by avoiding intermediaries and cultivating direct relations with consumers. He suggested, and producers seized on the idea of creating, a project to market high-quality organic beef. After three roller-coaster years of preparation, Ranchers' Choice began processing and marketing lean, organically grown kosher beef in Sanford (Conejos County). Despite attracting extensive national publicity and garnering outside technical and financial support, the plant closed within six months; good intentions had failed to overcome the classic missteps of beginning businesses. Meanwhile, Carter had his organization on record in support of organic farming as an alternative to conventional agriculture. Not unexpectedly, that irked the majority of his members who depended on chemical-intensive cultivation.[58]

Despite rapid and dramatic changes in agricultural techniques, policies and the values on which those policies were based continued to lag behind. To be sure, since the 1980s, more and more commodity and livestock producers have adapted practices pioneered by natural and organic agriculturists. In addition, urban growth and development has meant less land and less water for agriculture. At the same time, the ever-increasing demand for food locally grown has helped revitalize natural and organic agriculture. One continues to hope that enthusiasm for Colorado's healthy lifestyle will translate into the recognition, the will, and the imagination to achieve that "balance between man and nature" Governor Lamm and others have long advocated.

❦

7

Organic by Choice

The influx of urbanites to Colorado helped create what began as a niche market for locally grown natural and organic foods. Colorado's scenery and healthy lifestyle also attracted socially minded entrepreneurs who turned small-scale producers into multi-billion-dollar distribution businesses, often scaling up organic production and making the organic seal widely recognized by the general public. But there is more to organic agriculture than its still fractional part of overall agricultural production. More, too, than a technique, cultivating organically has become the most visible expression of a spiritual value, respect for the holy earth, a beacon of hope that Coloradoans will somehow find a way to live within the natural limits of the land and water.

In April 2014 the Rocky Mountain Conference of the United Methodist Church (UMC) awarded $5,000 to Hope UMC in Greenwood Village, a Denver suburb, to explore the possibility of using 9.7 acres of church-owned rangeland in soon-to-be developed northeast Aurora as the site for a new faith community that integrates organic agriculture and spiritual rituals. The goal was to answer the question, What does faithful living look like in the

https://doi.org/10.5876/9781646422050.c007

twenty-first century? The driving force behind this experiment was Stephanie L. Price, the passionate but unassuming youthful associate pastor of Hope UMC. Originally from rural upstate New York, Price had earned a master of divinity degree at the Iliff School of Theology in Denver. In 2012 Hope UMC congregant Bill Stevenson, director of the Rocky Mountain Farmers Union Cooperative Development Center, had contracted with Price to help build relationships between local churches and agricultural cooperatives, producers and consumers. In July 2018, Price became the full-time pastor of the new faith community, simply called The Land. But this occurred only after a team of Hope UMC congregants, supported by landscape architects and civil engineers, had drafted an implementation plan readily approved by the City of Aurora. The municipality agreed to provide water to the site for agricultural use, prior to development of surrounding neighborhoods that would require water for domestic use.[1]

To visitors to the site in fall 2018, Reverend Price envisioned a church without brick and mortar consisting of an edible labyrinth where congregants would grow food for neighbors in need, an organic orchard producing fruits and nuts, a farm stand to distribute food, and a cathedral greenhouse with indoor space to grow produce and gather people. At first blush, one might scoff at this experiment in agri-spirituality, casting it aside as eccentrically trivial. But following in the footsteps of the eighteenth-century English cleric John Wesley, founder of her denomination, Reverend Price seeks to instill a practical theology: recasting traditional Judeo-Christian ethics to fit contemporary circumstances, in belated recognition of nature's limits. By putting "service above sacrament, relationship above ritual," Price radiates hopeful confidence in the spirit of the Union Colony's utopian Nathan Meeker and the Rocky Mountain Farmers Union's idealistic Vince Shively.[2]

The spiritual dimension of caring for the land may have always existed, but the life-sustaining aspect—food production—has taken on a sense of urgency. From an economic point of view, farmers union President Dave Carter had envisioned organic farming as a way to save small and mid-size farmers while caring for the land. He had taken a leadership role on the ad hoc committee advising the Colorado Department of Agriculture on criteria for the state's organic certification program. After leaving the farmers union, he served as chair of the US Department of Agriculture (USDA) National Organic Standards Board during the period (2002–4) when rules to implement the

Organic Foods Production Act of 1990 were first put into place. Colorado certified its first two farms in 2002, with thirty more in 2003. Essentially, the act defined organic foods as foods started with organic seed or organic transplants and grown without the use of most conventional pesticides and herbicides, without fertilizers made with synthetic ingredients or sewage sludge, and without ionizing radiation or bioengineering. Organic meats, poultry, eggs, and dairy products are from animals that are given no antibiotics or growth hormones. Before a product can be labeled USDA organic, a USDA-approved certifier must inspect the farm or ranch where the product is grown to ensure that organic standards are met. Intermediaries who process and handle organic food must also be certified. The Colorado Department of Agriculture remains the principal federally accredited organic certifier and regulatory authority in Colorado and maintains a readily available online list of certified organic operations in the state.[3]

The USDA organic label assures consumers that they are purchasing products grown according to specific standards. But it suggests more than that. By placing emphasis on renewable resources, conserving land and water, and "management practices that restore, maintain, and enhance ecological balance," organic cultivation in effect keeps costs internal to the farm.[4] By reducing the need for expensive, energy-intensive inputs such as synthetic fertilizers, organic cultivation is limited to the farm by natural cycles, making it easy to track and putting cost control within the farmer's means. In contrast, conventional cultivation generally focuses on higher yields and generating direct and indirect costs external to the farm, which obscures true production costs, stretching beyond nature's limits.[5]

Often lost in the debate over conscientious stewardship of land and water is recognition of the continuities between organic methods and long-established conservation practices such as rotating crops, using green manures, and integrating crop and livestock production. A further misconception is that organic farming is somehow antithetical to new technology; in fact, the most successful organic farmers and ranchers use the latest energy-efficient machinery, sophisticated electronic equipment and software, the most site-adaptable plant cultivars and livestock breeds, and the best conservation practices to maintain and improve soil health. To be sure, tilling, a common practice in organic farming, causes some destruction of soil organic matter; but it can also turn under and mix vital manures and plant residues into the soil, increasing organic matter. This is especially beneficial to Colorado soils that are relatively low, ranging from 0.5 percent to 3 percent in soil organic matter. Tilling in manures and residues improves water filtration and retention, increases microbial activity, and slows the release of nutrients, to the benefit of plant growth.

In 2008 the USDA Agricultural Marketing Service, the agency administering the implementation of the organic act, published results of its first annual state-by-state survey of certified organic agriculture. Colorado reported 220 certified organic farms, 153,981 acres in certified organic farm and ranch land, and over $70 million in sales of certified organic products. In 2016 the survey reported 181 organic farms; 176,496 organic acres, of which 118,641 were in cropland; and $181 million in sales. The Colorado Department of Agriculture estimated that Colorado farmers, ranchers, and food processors accounted for roughly $2.5 billion in organic sales, or about 10 percent of the nation's sales of organic products. Colorado ranked first in organic wheat acreage, third in millet and lettuce, fourth in all cropland and all grain, and sixth in all vegetable acreage, including potatoes.[6]

Among the best-known Colorado pioneers in organic agriculture was Lewis O. Grant. Following military service during World War II, he worked in Israel's Negev Desert region, designing and installing state-of-the-art irrigation systems. He returned to the United States to earn a master's degree at California Institute of Technology and joined the Colorado State University (CSU) engineering faculty in 1959. Shortly thereafter, he and his son Andrew purchased 100 acres near Wellington—an area dominated by commodity crops and livestock production—to exclusively grow irrigated vegetables. In 1974 the Grants converted to organic only; their farm was one of the earliest certified under the state organic certification program, for which the Grants had strongly advocated.[7]

Grant Family Farms quickly expanded to over 2,000 acres, cultivating more than 100 different types of vegetables—mostly in the onion (Alliaceae), lettuce (Asteraceae), cabbage (Brassicaeae), spinach (Chenopodiaceae), and tomato/potato (Solanaceae) families—as well as pasture-raised, organically fed chicken. The Grants managed their operation with a system introduced from Europe known as Community-Supported Agriculture (CSA), under which growers and consumers share the risks and benefits of food production. In its heyday, with more than 5,000 consumer members, Grant Family

Farms was recognized as the nation's largest CSA. Members purchased shares of expected production in advance of each growing season; during harvest season, they collected their shares weekly at delivery locations throughout the region. The principal advantage of the CSA for consumers was the ability to purchase healthful foods locally grown; for farmers, the benefits included securing working capital in advance of purchasing seeds and supplies, enjoying a degree of financial security, earning better prices for their produce, and having an inexpensive way to market their products. As with many well-intentioned small businesses, Grant Family Farms grew too quickly; some missteps, unexpected risks, and just plain bad luck led to bankruptcy in 2013. Under new ownership, the farmland remains certified organic, today producing pinto beans and field corn.[8]

Berry Patch Farms in Brighton, a few miles northeast of Denver, represents another business design. Co-owners Tim and Claudia Ferrell bought the 40-acre farm in 1991, inaugurated a system in which consumers harvest their own produce, and obtained organic certification in 2003. Tim Ferrell explained that in recent years the organic label had taken on a greater meaning for consumers, but the Ferrells' decision to cultivate organically was fundamentally an ethical one. "We just do not feel comfortable using fertilizer that would infiltrate the water table," he told a reporter. As with most decisions, farming organically has occurred for a variety of reasons, depending on changing circumstances. J. I. Rodale's publications during the 1970s had at best only a marginal impact in convincing Colorado farmers to cultivate organically. Yet for reasons unclear, Gerald Monroe made the decision in 1936 to farm strictly organically. Monroe Organic Farms near Kersey (Weld County) has been considered the state's oldest organic farm by choice; it was among the first certified organic farms in Colorado and currently operates as a CSA.[9]

For large-size producers as well as for some processors, the transition to organic certified has been a strictly economic decision. Take Hungenberg Produce, a relative newcomer to certified organic farming. This family farm, started on 7 acres in 1908, now comprises 4,000 acres—planted primarily to carrots and cabbage—in the vicinity of Greeley. Responding to consumer demand, co-owner Jordan Hungenberg decided to experiment by converting 62 acres from conventional to organically grown carrots. In 2016 he began that transition, which generally requires three years in accordance with organic guidelines. Despite higher labor costs and a midsummer hailstorm, the future organic plat earned a profit during the first year, which encouraged Hungenberg to triple his organic acreage and become the largest supplier of organic carrots in the state. Similarly, processors of commodity crops are providing incentives. Denver-based Ardent Mills, the nation's leading supplier of flour, encourages grain farmers by offering to pay a premium price for crops grown on land in transition while the farmers work toward organic certification. Ardent's principal partners are Cargill and ConAgra, the same corporations heavily invested in the production of grains in Colorado.[10]

The highly visible appeal of locally grown organic vegetables and fruits has tended to obscure significant strides toward organic cultivation of commodity crops. A case in point: the cultivation of certified organic millet initiated by the Hediger family. Indeed, Jean Hediger was among the small group that included Mel Coleman Sr., Lewis and Andrew Grant, and the farmers union's Dave Carter in helping draft Colorado's first organic certification regulations. From a small farm near Fort Collins, the Hediger family farm grew to 2,500 dry-land acres centered at Nunn (Weld County). Instead of cultivating a single crop such as wheat, the family began experimenting with millet and oats, also planting yellow clover to replenish the soil with a natural source of nitrogen. They discovered that millet (the "golden grain") is especially well adapted to Colorado conditions, growing without supplemental irrigation while meeting expanding consumer demand for gluten-free grains. In recent years, the Hedigers established the Golden Prairie label to market and distribute organic grains and organized a millet growers' cooperative of more than a dozen farmers cultivating 40,000 acres. The Hedigers continue to embrace new technologies in both cultivation, such as the use of unmanned aerial vehicles (drones) to monitor their fields, and ever-changing approaches to marketing. Through wholesalers and distributors such as United Natural Foods, Inc., Golden Prairie products reach consumers nationwide. Indeed, Colorado ranks first in the nation for the production of organic millet.[11]

Perhaps more than any other two individuals, Boulder-based entrepreneurs Paul Repetto and Mark A. Retzloff have been credited with scaling up organic production into billion-dollar corporations. Perhaps their best-known brand is Horizon Organic Dairy, which they co-founded in Boulder as Horizon Natural in 1991; they expanded distribution nationwide and converted their private company into a publicly traded company in 1998. Since then, the company has gone through several stages of ownership. Horizon Natural was acquired in 2017 by Danone, a French multinational food products company, and took the name DanoneWave, after the Broomfield-based holding company White Wave that sold the company to Danone. Through contracts with hundreds of organic-certified dairy farms, Horizon Organic is the nation's single largest supplier of organic milk and a variety of other organic products.[12]

Retzloff in particular has successfully managed to revitalize a number of natural and organic food companies. After earning a degree in environmental studies, conservation, and resource planning at the University of Michigan, he transformed a natural foods cooperative in Ann Arbor into Eden Foods, still a major independent producer of organic foods. After moving to Boulder, he assisted in converting Pearl Street Market, a local natural foods store, into a regional chain that was renamed Alfalfa's—which in the 1990s merged with Wild Oats Market, later acquired by Whole Foods Market. After twenty-five years, Alfalfa's has been reestablished, with several stores in northern Colorado. Meanwhile, Retzloff managed to turn around Rudi's Organic Bakery, which had started in 1976 as Rudi's Bakery, a small Boulder bakery specializing in dark whole grain breads considered higher in nutritional value than white breads. Rudi's Organic Bakery is now a large state-of-the-art bakery in downtown Boulder, one of nearly fifty brands belonging to the Hain Celestial Group.[13]

As with fast-growing corporations in other areas of the economy, the large producers of organic products experienced their share of travails. This was most apparent with the dairies, in part because dairy products are the most popular organic products. From time to time, Boulder-based companies Horizon Organic and Aurora Organic were criticized by consumer groups for failure to adhere to National Organic Program standards; for example, the perception held that they were not having cows graze on pastureland during growing seasons. The standards require that, weather permitting, cows must graze at least 120 days per year and obtain at least 30 percent of their dry-matter food by grazing on organic pastures (hay or grain). Both companies have been in compliance for at least a decade. Founded as a small dairy in Jerome, Idaho, in 1976, Aurora Dairy built its first dairy operation in Colorado in 1980, then moved its headquarters to Boulder four years later. Mark Retzloff helped manage the company's transition to Aurora Organic Dairy, which became 100 percent certified organic in 2003. At present, Aurora Organic operates two large certified organic farms, each with thousands of dairy cows, near Platteville and Gill in Weld County, as well as two large farms and a milk plant in Texas and one in Missouri. All of these plants produce and package private-label milk and other dairy products for major chains, among them Costco, Safeway, Target, and Walmart. As a certified Benefit Corporation (B-Corp), Aurora Organic is legally bound, first and foremost, to provide societal and environmental benefit to society and, secondarily, to provide profits for its investors. Since organic milk and dairy products are considered "gateway products" for attracting other organic purchases, Aurora Organic has engaged in a deliberate effort to encourage more and more consumers to buy organic products.[14]

Boulder County has become home to a growing number of regional, national, and global brands of natural and organic foods. In 2005 a group of industry leaders and other interested individuals established Naturally Boulder as their business association, serving as a common voice as well as a facilitator for networking, mentoring, and educational events and activities. By using the word *natural*, Naturally Boulder does not limit itself to organics; instead, the company more generally promotes those products and practices "made with the goal of promoting human, animal and environmental health" and opposes those that are or are perceived to be contrary to the "Boulder brand, which stands for a healthy lifestyle." The association claims about 1,000 industry members and 100 supporting businesses, nonprofits, and government organizations.[15]

In 2011 Naturally Boulder entered the local public debate on genetically modified seeds, better known as genetically modified organisms (GMOs); at issue is whether GMOs would be permitted on several thousand acres of county-managed agricultural lands, part of the county's open space initiative. Ten years earlier, in response to concerns about possible pollen drift, an informal group of local farmers and homeowners persuaded the county commissioners to establish a GMO technical advisory committee, specifically to recommend limits, if any, on the use of genetically modified corn seeds on county-managed land. Committee members heard, on the one hand, from mostly part-time organic farmers and natural food enthusiasts and, on the other hand, from conventional farmers and their production

FIGURE 7.1. Agronomist and Colorado State University students evaluating pasture grasses, 2019. *Courtesy*, Aurora Organic Dairy, Gill, CO.

agriculture supporters. Organic and natural food supporters made up the overwhelming majority of those present at the hearings. The commissioners, representing the longer-standing powers that be, voted against the GMO ban, basing their decision on evidence presented by county staff. This very public controversy attracted considerable national attention.[16]

The controversy continued. By the time Naturally Boulder voiced its opposition to GMOs, 86 percent of all corn, 93 percent of all soy, and 95 percent of all sugar beets grown in the United States were from genetically modified seed. The association noted that this state of affairs left few choices for farmers and processors who sought to avoid GMOs and argued that GMO seeds had yet to be in production long enough to determine their safe use. In 2016 a newly elected group of county commissioners voted to ban genetically engineered crops on county-managed land. At the time, farmers on that land were planting from 400 acres to 600 acres in GMO sugar beets seeds and 1,500 acres to 1,800 acres in GMO corn kernels. The new commission announced that those farmers were required to complete their transition to non-GMO seeds by 2025. General opposition from conventional farmers and their suppliers, combined with the confusion brought on by the controversial decision, caused commissioners to consider delaying their deadline.[17]

Unquestionably, bioengineering is an effective way to modify the genetics of seeds for specific farming techniques and soil conditions, thereby increasing yields and helping to keep food prices low for consumers. Bioengineering does raise questions about extending the limits of nature, possibly threatening the integrity of the ecosystem and perhaps even generating societal costs yet to be understood. Public debate on the matter has tended to mix ideology and sentimentality with science and empirical data. It is often the case that we can better understand an idea, practice, or system by defining what it is not: all techniques of bioengineering are disallowed in organic production because they could not occur under natural conditions. Genetically modifying seed is simply the best-known among those techniques.

At the risk of oversimplification, genetic engineering means transferring a gene—a unit of inheritance located on a chromosome and the DNA within the nucleus of a cell—from one organism to another, thereby altering the characteristics of the recipient organism. In genetic engineering, the technician modifies the genetic material of organisms, whether cisgenic (related) or transgenic (unrelated), radically changing its evolution. By contrast, the farmer or technician engaged in hybridization breeds two related organisms through traditional approaches in favor of certain characteristic by controlling sexual reproduction, thereby expediting selection. Promoters of genetically modified crops would argue that the genes inserted through bioengineering are natural, that is, products of nature. Therefore, transgenic organisms are no different from their non-genetically engineered counterparts.[18]

Among the early techniques of genetic modification was the use of a device known as a gene gun, which propels microscopic gold balls coated with the DNA for a gene from one organism to another organism of a different species. Many microscopic balls miss their target, but those that reach it deliver the sender's DNA into the receiver's DNA. The modified plant then becomes tolerant of certain herbicides, which reduces the need to control weeds by means of cultivating but increases reliance on herbicides such as glyphosate. Adding to the complexities of early transgenic manipulation, agricultural scientists have developed procedures to insert genes for antibiotic resistance, raising concerns among public health officials about the overuse or misuse of those critical agents. The combination of proprietary plant genetics, increased herbicide usage, and a potential decline in antibiotic usefulness helps explain why the original organic rules disallowed breeding processes not possible under natural conditions and still do today. In addition, the semi-random nature of this technique makes its replication time-consuming, expensive, and—most limiting—imprecise.[19]

The insertion of genetic material from one organism to another, however, does occur in nature. Crown gall is a plant disease caused by a soil-borne bacterium (*Agrobacterium tumefaciens*, also named *Rhizobium radiobacter*), which occurs on many different plant species. An infected plant exhibits tumor-like lesions that consume the plant's resources. Occupying space between plant cells, the bacteria can transfer a small segment of their own DNA into the plant's genetic material (genome). That transgenic transfer causes the plant to grow in unusual, self-destructive ways that eventually lead to the breaking open of the plant's cellular structures, causing its death and releasing the causal bacteria back into the soil to infect again. By understanding this process, scientists can use it to insert a gene from one organism into another, thereby creating genetically modified organisms. This technique is much more precise than the gene gun, but the success rate is still relatively low for some species.[20]

The GMO ban imposed by the Boulder County commissioners on lands managed by the county, combined with the negative reaction by commodity producers and agri-businesses, puts Boulder at the center of the debate over ecologically acceptable agricultural practices—more generally, over what makes a healthy, dynamic, and lasting community and, by extension, a livable world. Observing this debate with keen interest is Andrew Hopp, an independent sales representative covering Boulder County for DuPont Pioneer, successor to the Pioneer Hi-Bred Corn Company founded by Henry A. Wallace in 1926. A fifth-generation member of a Larimer-Weld County farm family, Andrew is a 2018 CSU graduate in soil and crop science with a minor in organic agriculture because he wanted a broad understanding of the prospects and limits of agricultural technology. His principal customers are sugar beet and field corn farmers. More than 95 percent of sugar beet seeds are genetically modified so the crops can tolerate the herbicide glyphosate (Roundup). There is no question that herbicide-tolerant crops contribute some environmental benefits, facilitating conservation tillage and reduction in fuel use. But as agricultural scientists had predicted, the use and misuse of Roundup has resulted in the emergence of glyphosate-resistant weeds, which has forced the chemical industry to develop further modifications to crop seeds. Glyphosate also poses potential risks to human health, which has generated thousands of lawsuits against the herbicide's manufacturers.[21]

Recognizing that the development of genetically engineered seeds has been regulated by a coordinated effort of three federal agencies—the US Department of Agriculture (USDA), Environmental Protection Agency (EPA), and Food and Drug Administration (FDA)—Andrew Hopp believes conventional farmers have been timid in defending their use. At first, GMOs frightened all of them, he noted, so they approached the matter with great caution. Beyond the objections from ideologues who are simply "anti-GMO," he added that agriculturists have come to appreciate the limits of that technology. For example, field corn commonly carries a genetically engineered trait that produces insecticidal proteins from a soil bacterium (*Bacillus thuringiensis*, or Bt), which is consumed by larvae of corn rootworm, corn earworm, and European corn borer—causing the death of the pest if it feeds on roots or shoots. Often, that trait is incorporated, or "trait stacked," with another genetically modified trait, most commonly glyphosate resistance. The EPA now allows DuPont Pioneer to continue using the Bt trait aboveground but not belowground in the roots because scientists discovered that belowground pests have developed resistance.[22]

With regard to producing healthful and affordable foods while conserving natural resources, Hopp holds more hope for gene editing than for gene modification. Developed about a decade ago, gene editing is a technique based on a stunning discovery: at some point in the distant past, a microscopic single-cell organism (prokaryote) that lacked a membrane-bound nucleus and other cell structures encountered a dangerous virus. In the process of combating those viruses, specific strands of the viral DNA became enmeshed with the prokaryote DNA. Those same strands exhibited the unique characteristic of repeating themselves and passing the "genetic memory" of the encounter from generation to generation, so that when another viral DNA encounter took place, the single-cell organism could detect and destroy ("snip off") the undesirable virus DNA strand. That process of repetition has been given the technical name clustered-regularly-interspaced-short-palindromic (double-stranded) repeats, commonly referred to as CRISPR.[23]

Applied to agriculture, gene editing can permanently remove undesirable traits from plants and animals. Using the example of field corn, gene editing could "snip off" the gene that attracts the corn borer caterpillar or even the

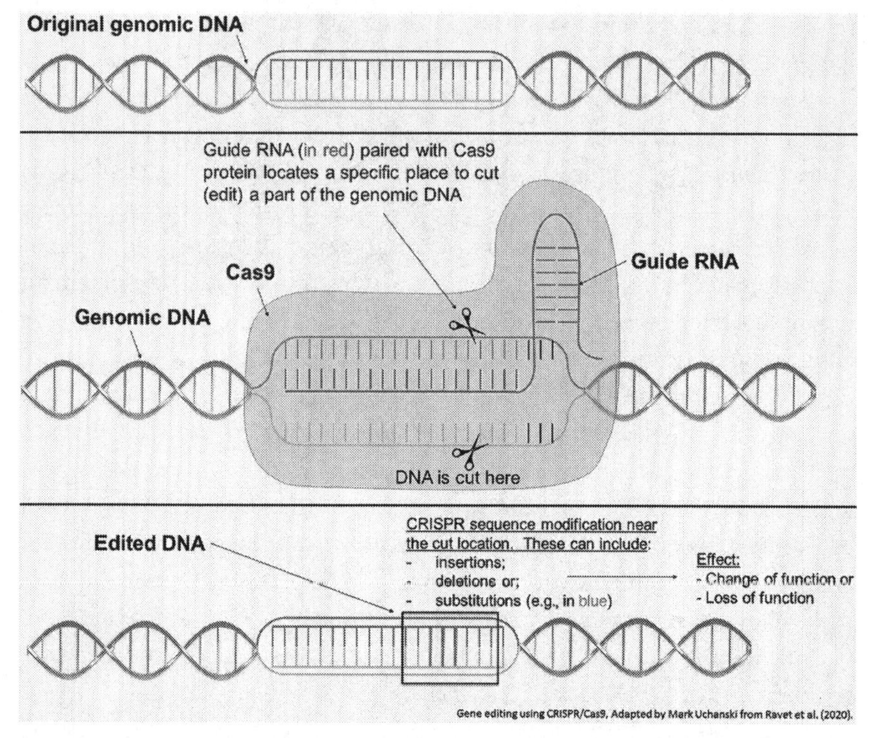

FIGURE 7.2. Gene editing using CRISPR/CAS9, adapted by Mark Uchanski. *Courtesy*, Karl Ravet and Stephen Pearce.

gene that creates the odor the caterpillar uses to find the corn plant in the first place. Unlike a specific gene inserted from one plant into another, which can be detected through simple DNA extraction and testing, gene editing is much more difficult to trace and is heritable, in some case causing permanent change for generations. Because gene editing applies to humans as well, it takes very little to imagine the profound ethical questions civil society has yet to address.

In 2019 the agricultural biotechnology company Calyxt contracted with soybean farmers on 34,000 acres in the upper Midwest to plant the first commercially available gene-edited seeds designed for consumer health, to be used in the production of Calyno™, a vegetable oil for cooking and salad dressing with 20 percent less saturated fatty acids than non-engineered soybeans and zero grams of trans-fatty acids. A year earlier, Secretary of Agriculture Sonny Perdue had announced that unlike GMOs that were highly regulated, the USDA had no plans to regulate new plant cultivars that result from gene editing. In contrast, the European Union put gene-edited seeds in the same category as genetically engineered seeds, thus prohibiting their use among member nations. Earlier, in 2016, the board of the USDA National Organic Program revised its Rule 205.105 to specifically exclude genetic engineering and genetic editing (CRISPR) as allowable methods for organic production; neither can be tracked like genetically modified organisms can.[24]

At least in the short term, bioengineered crops hold promise for strengthening the food supply and food quality by increasing yields, lowering production costs, and presumably lowering costs to consumers; they also contribute to environmental protection by requiring that the soil be tilled less and by using fewer chemicals. Bioengineering techniques are relatively new, however, so questions remain as to the safety of certain gene-edited crops for both humans and the environment. Ethical implications have yet to be critically and broadly examined. The public is becoming increasingly aware of the presence of bioengineered crops in the food supply, which helps explain growing interest in the production and processing of natural and organic vegetables and fruits.

Colorado agriculture has long been known for its specialty crops as well as for its commodity crops and livestock production. *Specialty crop* is the generic term the USDA uses for "fruits and vegetables, tree nuts, dried fruits and horticulture and nursery crops, including floriculture." Specialty crops are not commodity crops. Consumers share the widely held belief that fruits and vegetables grown at Colorado's higher elevations under semiarid conditions are superior to produce grown in humid surroundings at lower elevations. Unquestionably, certain specialty crops perform well in Colorado's climate, topography and, yes, its demography. The burgeoning demand for locally grown healthful food reflects the postwar interest in living the healthy Colorado lifestyle. Moreover, specialty crops help connect urban consumers and rural producers, thereby contributing to better public understanding of agriculture, its costs, and its benefits.[25]

In 2002, with financial support from the USDA, the CSU College of Agriculture in partnership with the Colorado Department of Agriculture inaugurated a specialty crops program home-based in the Department of Horticulture and Landscape Architecture. Frank Stonaker was appointed to develop and manage the new program, succeeded in late 2015 by coauthor Mark Uchanski. Stonaker's father, Howard H., had led the Fort Lewis

(Hesperus) cattle-breeding project (1946–66) and became an internationally recognized authority on the subject. The family had bought a small farm east of Fort Collins, where Frank worked until the mid-1980s, when the property was sold to Thornton as part of the city's purchase of farmlands for its water rights. After earning an undergraduate degree in agronomy at CSU and a graduate degree in entomology (biological pest control) at the University of Florida, Stonaker returned to farming, including work in organic greenhouses and as production manager at Grant Family Farms.[26]

Once back at CSU, Stonaker completed the transition from conventional to organic cultivation at the horticulture farm northeast of Fort Collins, which was part of the experiment station, by then renamed the Agricultural Research, Development and Education Center (ARDEC). Since 2002, the specialty crops program has captured the interest of non-farm undergraduates from a multitude of academic majors, all intrigued by growing healthful vegetables and fruits and by the commercial potential of high-value farming in and around urban centers. As a result, CSU instituted an interdisciplinary minor in organic agriculture, which has attracted organic industry support—most notably from Aurora Organic Dairy in the form of scholarships, teaching experiences in organic agriculture, and experiential learning opportunities. During Stonaker's tenure, the horticulture farm expanded to include seven season-extending high tunnels, unheated plastic-covered structures also known as hoop houses, and 11 acres of certified organic open field space.[27]

As part of the university's extension efforts, Frank Stonaker obtained USDA support to launch the Rocky Mountain Small Organic Farm Project, which provided four to six competitive grants annually (a total of sixty-five between 2002 and 2013) for growers to participate in training and education activities as well as to engage in experimentation on their own farms. Taking what the growers learned, combined with his own research at the experiment station, Stonaker prepared "A Guide for Small-Scale Organic Vegetable Farmers in the Rocky Mountain Region," arguably the best summary of the variety of Colorado's growing conditions and the cultivation techniques needed to ensure optimal production of organics; it includes an annotated list of vegetables that grow especially well in the region. Stonaker's "Guide" superseded the standard reference manual for small-scale vegetable growing published more than seventy years ago by researchers at the Cheyenne Horticultural Field Station. In 2013, Stonaker left CSU and moved to Hotchkiss where he purchased his own farmstead, which included a 30-acre irrigated orchard (organic peaches, apples, grapes) and a 1-acre plot of organic sweet cherries grown under high tunnels. He interplanted the orchard with vegetables as a way to control weeds, retain soil moisture, and eliminate bare ground. When CSU reopened the Rogers Mesa (Hotchkiss) experimental substation in 2018, the university brought Stonaker back to manage the facility, overseeing the transition to all certified organic and providing advice to individual farmers contemplating their own transition to organics.[28]

Among the early participants in the Stonaker small organic farm project was Daniel G. Hobbs, part of a new generation of mostly young people without farm roots but deeply committed to the ethical stewardship of the land and to the cooperative approach in developing competitive economies of scale to produce healthful foods locally. A fifth-generation Denver native, Hobbs had been privileged to spend his high school summers working in Ecuador and Paraguay, where he became fascinated by small-scale farming. He started college at Earlham, Indiana, living and learning on the college farm. After graduating with a liberal arts degree from the University of New Mexico, Hobbs went to work for NewFarms Agricultural Services, a charitable nonprofit helping mostly subsistence farmers in northern New Mexico restore and improve agricultural lands and market their produce. As an outgrowth of that work, he became general manager of Tres Rios Agricultural Cooperative, a producer-owned marketing and distribution co-op reaching into southern Colorado, which led him to purchase his own 30-acre irrigated farm in Avondale, east of Pueblo. Meanwhile, Robert R. Mailander, the first manager of the Rocky Mountain Farmers Union cooperative development center, had contracted with Hobbs to assist in creating year-round farmers markets. In 2002, Hobbs was selected to participate in the CSU organic farm project, which enabled him to firmly establish Hobbs Family Farms, where he continues to grow certified organic garlic, open pollinated seeds for national seed companies, fresh vegetables, and field crops for local markets.[29]

Dan Hobbs has continued to work with small to mid-sized farmers through a variety of cooperative ventures, with the goal of aggregating produce to better compete in the market. In the lower Arkansas Valley, he led the effort to establish Excelsior Farmers Exchange. Located in a decommissioned schoolhouse in Boone, not far from his farm, the exchange is organized as a "food hub" to collect, store, process, market, and distribute food products to grocers in Pueblo and Colorado Springs. Hobbs envisions the exchange as a place that encourages natural and organic farming techniques, creates year-round jobs through value-added processing, and helps grower members better cope with recurring drought and diminishing agricultural waters due to diversion to the cities. Especially for small-scale farmers operating in parts of the state distant from urban markets, Hobbs considers food hubs to be marketing vehicles that are superior to farmers markets; best of all, they facilitate purchasing by institutions through programs such as farm-to-school that are especially favorable to organic farmers, ranchers, and dairy operators since the program's intent is to purchase locally grown fresh foods such as vegetables and fruits, eggs, and meats. Hobbs believes small and mid-sized growers are best positioned to capitalize on the marketing of the fruits and vegetables for which Colorado is best known—Rocky Ford melons, Pueblo chiles, Palisade peaches, and Olathe sweet corn, among others. The principal site for experimentation on such specialty products as melons and chiles is CSU's Arkansas Valley Research Center in nearby Rocky Ford.[30]

Hobbs regards Michael Bartolo, the station's chief research scientist, as his mentor in the agricultural sciences. Bartolo comes from a line of farmers. His grandparents emigrated from Italy to the United States, settling on St. Charles Mesa near Pueblo; they brought chile seeds that after more than a century of the family's seed selection and seed saving, adapted well to growing conditions in the Arkansas Valley. After earning undergraduate and master's degrees in horticulture at CSU and a PhD in plant physiology at the University of Minnesota, Bartolo returned to Colorado and began working at the Rocky Ford Research Center. Taking seeds from chile plants passed down through his family, Bartolo selected further, developing several new cultivars, with the most popular being 'Mosco.' Most of his research, however, has been on onions, Colorado's principal high-value vegetable crop.[31]

Melon growers have credited Bartolo with saving their businesses. In 2011 producers in the Rocky Ford area experienced a nearly catastrophic blow. Thirty-three people died and another 150 were seriously sickened by eating melon infected by the bacterium *Listeria monocytogenes*, which was traced to a packing facility in Holly (Prowers County), about 90 miles east of Rocky Ford. Growers had already been considering making changes in packing methods; the 2011 outbreak jolted them into action. Bartolo credits Rocky Ford grower Michael Hirakata for leading the effort to restore the Rocky

Ford cantaloupe's well-earned reputation and prevent such an outbreak from happening again. Hirakata was in the process of planning a new cantaloupe facility on his farm, started by his family in 1915; adverse publicity about the outbreak attracted food safety experts from across the nation willing to provide advice on specifications for a new facility. In the past, some farmers had washed melons in recycled water, failed to add the proper amount of chlorine, and used conveyors and wash basins that were difficult to keep clean. Since the outbreak, Hirakata has been using fresh water with the proper amount of chlorine and plastic rollers with specially designed spray nozzles. In addition to bringing in experts to assist farmers and packers, Bartolo helped organize the Rocky Ford Growers Association, which adopted new food safety standards and sponsored a public relations campaign to reestablish the Rocky Ford melon's image.[32]

Growing interest in safe and healthful foods combined with the rising political influence of the natural and organic foods industry translated into a big boost for fruit and vegetable growers. The Farm Bill of 2014 passed the US Congress with strong bipartisan support and was signed by President Barack Obama, further reflecting the shift in consumer values and thus in agricultural practices. The bill raised support for specialty crops to $3 billion, a 50 percent increase over the Farm Bill of 2008; reduced subsidies for commodity crops to $23 billion, a 30 percent decrease; and increased funding for conservation projects. Funds for state specialty crop programs established to promote production and increase competitiveness rose from $52 million to $72.5 million. Funding for specialty crop research doubled, to $80 million; the amount for farmers market and local food promotion tripled, to $30 million. Most notable, however, the Farm Bill of 2014 increased funding for the National Organic Program and reinforced its authority to protect the integrity of the USDA organic label, thereby strengthening consumer confidence and justifying higher prices that benefited certified organic producers. The bill also exempted organically grown commodity crops from the usual commodity "check-offs" and even enabled them to devise their own promotional organic "check-off." Most significant financially, the bill directed improvements in crop insurance, enabling new organic farmers to purchase coverage that in case of losses would pay actual market prices—more favorable reimbursement than that for conventional crop coverage. To encourage the transition from conventional to organic farming, Congress established an organic cost-share program, enabling the USDA to reimburse producers up to $750 annually for the certification process through state departments of agriculture.[33]

By loosening restrictions on the production of industrial hemp, the 2014 Farm Bill opened up a yet to be fully understood opportunity for Colorado agriculturists. For millennia, industrial hemp has been grown, most notably in China and France, and then spun into cloth fiber and rope. Modern techniques have been developed to refine industrial hemp into a variety of products, including oils, pulp for making paper, biofuels, and biodegradable plastics. For more than seventy-five years, federal regulators had not distinguished between industrial hemp and marijuana. Industrial hemp includes a distinct non-psychoactive set of cultivars of *Cannabis sativa* L. with a THC (tetrahydrocannabinols) concentration of not more than 0.3 percent; marijuana is a psychoactive set of cultivars with THC levels in excess of 15 percent by weight. In 2012 Colorado voters approved Amendment 64 to legalize marijuana for recreational as well as medicinal purposes. The first recreational marijuana stores opened two years later. In a joint statement, the USDA, Drug Enforcement Agency, and FDA expressed the view that the Farm Bill legalized the cultivation of industrial hemp for research purposes where allowed by state laws. The federal statute specifically limited such cultivation to institutions of higher learning or state departments of agriculture "for purposes of agricultural or other academic research or under the auspices of a state agricultural pilot program for the growth, cultivation or marketing of industrial hemp."[34]

Confusion about the non-psychoactive and psychoactive types of cannabis that cannot be distinguished visually contributed to interest in cultivating industrial hemp. In response to inquiries from businesses and the general public, scientists at CSU's Southwestern Colorado Research Center near Cortez began the first industrial hemp cultivar trials of seeding rates, fertility, and deficit irrigation—allowing plants to withstand mild water stress. More trials have occurred each year since 2015, expanded to other CSU research centers. Legalization of the production and sale of hemp, part of the 2018 Farm Bill, precipitated a rush to get into the business—overwhelming extension offices with requests for information and requiring staff members to inform themselves on cultivation practices and the process for registering with the Colorado Department of Agriculture. By spring 2019, more than

1,000 hemp farms had registered with the department, although it remains to be seen if industrial hemp is a viable crop in Colorado.[35]

Meanwhile, for growers seeking high profits with relatively little labor and for counties and municipalities desperate for new revenues, the legalization of marijuana and industrial hemp has added to the competition over use of agricultural lands. In Pueblo County, officials envision replacing chile and pastureland by establishing "the Napa Valley of legal weed." In the San Luis Valley, where family farms and small towns continue to struggle economically, plans are under way to both grow and process marijuana.[36] Yet to be fully understood—as with the case of poppies in Afghanistan—are the health, social, and economic costs.

Although it is too early to make judgments based on empirical evidence, the potential impact of the cannabis industry on Colorado agriculture and the overall environment cannot be ignored. Toward the end of his final term, Governor John Hickenlooper (D, 2011–19) expressed doubts about the legalization of marijuana and, given rising crime rates, whether marijuana should be recriminalized. His successor, Jared Polis (D), supports continued legalization and has assigned the commissioner of agriculture to navigate the uncharted waters.

Governor Polis selected Kate Greenberg, who was residing in Durango and serving as western states program director for the National Young Farmers Coalition, as the first woman to hold the position of commissioner. A native of Minnesota, Greenberg is a 2009 Whitman College graduate in environmental studies and humanities. Since moving to Colorado, she has worked closely with the farmers union, sharing the view that young people—especially those from non-farm backgrounds—are attracted to natural and organic farming and ranching and that non-conventional agriculture can provide an economically viable opportunity for small and mid-scale farmers to thrive. Shortly after entering office, Greenberg shifted personnel and financial resources into the agency's organic certification program—enabling more timely review of applications, more inspections of organic farms for compliance with national standards, and early participation in an "organic integrity database" of certified organic producers, processors, and distributors, which is accessible electronically to the general public.[37]

In May 2019, Governor Polis outlined the steps his administration intended to take to put Colorado on the path to achieve 100 percent reliance on renewable energy by 2040. He acknowledged Xcel Energy for its activities to reduce use of fossil fuels and increase use of wind and solar energy. Among the more unusual agriculture-related projects, in December 2018 Xcel Energy approved Jack's Solar Garden, LLC as a third-party provider of solar energy to its customers in Boulder and Weld Counties. The plan: once customers subscribe directly to the solar garden, they will receive credits on their monthly electric bill for the solar energy their subscriptions add to the Xcel Energy grid. Jack's Solar Garden may not be the typical participant in one of the nation's largest community solar programs. It has partnered with the US Department of Energy's National Renewable Energy Laboratory to build a 1.2-megawatt solar garden, where research scientists from the laboratory, Colorado State University, and the University of Arizona study how vegetables grown under solar panels contribute to soil and water conservation, at the same time possibly enabling solar panels to absorb energy more efficiently.[38]

The concept of farming in which vegetable crops share the land with solar panels is not entirely new. In the first decade of the twenty-first century, Japanese scientists began experimenting with agrivoltaic farming. In 2010, scientists from the French National Institute of Agronomic Research established a 9,000-square-foot agrivoltaic plot in an open field near Montpellier. Among their early discoveries: lettuce plants, which naturally tolerate partial shade, adapted to less light under the solar panels by growing larger leaves, and they did so in such a way as to absorb light more efficiently; the solar panels reduced evaporation and saved on irrigation water.[39] But Jack's Solar Garden is somewhat more ambitious. Named after owner Byron Kominek's grandfather, who bought the 24-acre farm between Longmont and Niwot in 1972, the garden has been designated as a solar garden research station by the Boulder County commissioners. Once fully operational, it will consist of 3,000 solar panels, generating enough electricity to power 600 homes; protecting crops from high temperatures, destructive hail, and late spring and early fall frosts; and saving on irrigation water. Before moving to the farm in 2016, Kominek had served in the US diplomatic corps in Mozambique and Zambia, where non-governmental organizations were engaged in helping to harness renewable energy for communities with limited resources.[40]

At the CSU horticulture farm near the campus, meanwhile, faculty and students in partnership with Sandbox Solar, a local solar panel installation

FIGURE 7.3. Hort farm, 2020. *Courtesy*, Agricultural Research, Development and Education Center, Colorado State University, Fort Collins.

company, started their own project in summer 2019. They placed nine sets of solar panels mounted on posts, which allowed the panels to be tilted and also raised from 4 feet to 9 feet aboveground to facilitate experiments with a diversity of organic crops as well as to allow a tractor to safely cultivate under the panels. Unlike Jack's Solar Garden, the CSU researchers are focused less on electricity generation capacity and more on balancing crop growth, shading, water use and distribution, protection from hail, and other commercial applications—with particular interest in new types of panels that allow more light to penetrate down to crops below.[41]

As Coloradoans hunger for locally grown healthy food, the Colorado Department of Agriculture has undertaken several initiatives, including the popular Colorado Proud marketing campaign. Grocery stores, farmers markets, restaurants, and garden centers use the Colorado Proud logo on locally grown, raised, and processed agricultural products—informing consumers about their purchases with the intention of steering more business to Colorado farmers, ranchers, and the flourishing "green" industry. The green industry contributes to making communities more livable and helps save water, especially considering that nearly half of all domestic water use has been for irrigating lawns.

Direct sales of Colorado food products from farmers to consumers have increased dramatically statewide over the past few years, to more than 100 farmers markets, 140 Community-Supported Agriculture programs, and 42 school districts (857 schools) participating in farm-to-school programs.[42] But access to fresh, healthful foods has not always been equitable or accessible, with low-income residents far more likely to suffer the consequences. Nearly one in six Colorado households and one in five children experience hunger or are food insecure, that is, lacking reliable access to a sufficient quantity of affordable, nutritious food; nearly 70 percent of Denver public school students are eligible for free and reduced lunches; and one-third of Denver families consume less than one serving of fruits and vegetables daily. To address this problem, Mayor Michael J. Hancock appointed Blake Angelo as Denver's first manager of food system development in June 2015. A member of the youthful generation of pragmatic idealists, Angelo graduated from CSU with degrees in agricultural economics; through extension, he served as CSU's first urban agriculture specialist. Working under the mayor, Angelo was the driving force behind the development of a "city-wide fresh food strategy" involving community gardens, urban farms, farmers markets, communal commercial kitchens, public markets, food hubs, mobile farm stands, neighborhood cooperatives, and food industry incubators. In 2017, Mayor Hancock announced Denver Food Vision and in 2018 the Denver Food Action Plan, which included among several goals that by 2030, at least 25 percent of all food purchased by public institutions would come from Colorado producers.[43]

A focus on imaginative urban and peri-urban agriculture does tend to minimize the far more prominent position of commodity and livestock production in Colorado and to underestimate the economic, social, and ecological value of all the goods and services that move agricultural products along the path from producer to consumer or final purchaser. By trying to measure the monetary value of each step along that path, agricultural economists at CSU concluded that agriculture ranks among the top three sectors of Colorado's economy rather than in the top ten normally considered. With the state's vast rangelands, livestock production remains the principal source of agricultural income, with the largest share of commodity crop production consumed by livestock.[44] Regardless of where one stands on the consumer preference or ideological spectrum, meat protein costs more to produce than vegetable protein; much of the food value of grains is lost when they are fed first to animals and only after that consumed as meat by humans. A beef steer requires about 20 pounds of feed to produce 1 pound of meat, compared to 10 pounds for a hog and 4 pounds for a chicken; in addition, a steer requires far more water—between 3 gallons and 30 gallons per day depending on ambient temperature—which takes us back to organic and natural foods. Health and healthcare professionals recommend that we eat smaller portions of meats and, when we do, preferably from grass-fed bison, cattle, pork, and lamb free from the artificial food supplements or antibiotics required in some processing operations. They also note that plant-based proteins (from legumes and pulses) contribute to human as well as soil health.[45]

No one is suggesting that organic is the only way to cultivate or that organically grown produce and livestock should be the only choice to feed the world, either now or in the foreseeable future. Farsighted agriculturalists note with approval the current situation in which the difference between conventional and organic systems is becoming less and less clear-cut, with conventional farmers adopting organic practices and organic farmers using appropriate technology developed for conventional farming. As more producers and consumers recognize the benefits of organics in improving soil health and addressing climate and demographic changes, the conviction that human ingenuity can outsmart Mother Nature begins to lose its luster.

The 1990 Farm Bill outlined a way to measure future contributions of organics to overall agricultural sustainability. In 2010 the National Academy of Science expanded on that effort with publication of a report titled *Toward Sustainable Agricultural Systems in the 21st Century*. The report further defined the four generally agreed-upon goals of sustainable agriculture contained in the bill: satisfy human food and fiber needs, enhance environmental quality, sustain the economic viability of agriculture, and improve the quality of life for agriculturists and society in general.[46]

In Colorado as in the rest of the nation, organic agriculture still occupies between 0.5 percent and 1 percent of cultivated land, generating about $180 million in sales annually (ranked ninth among US states) and growing; but this amount is still a small fraction of gross revenues from agricultural products. Recently, Denver charitable nonprofit groups such as Re:Vision have been assisting residents in low-income neighborhoods in converting their backyards into organic vegetable gardens, part of the city's effort to ensure access to healthful foods for all its residents. In addition, seeking to reduce food waste—estimated at 40 percent of food in the city—groups such as We Don't Waste help fulfill Denver's food needs by collecting and redistributing perishable items including fresh produce, lean protein foods, and dairy products.[47]

Of all cultivation methods known to date, organic is the closest to mimicking nature, which makes it the most environmentally friendly. In recent years, the term *regenerative agriculture* has come into vogue, describing a subset of organic agriculture that goes beyond maintaining soil health to enriching the soil, improving water retention, and increasing biodiversity among the plants, animals, insects, and microbes that make up a local ecosystem. Denver Botanic Gardens Chatfield Farms, located on 700 acres in the southwest metro area, devotes space to experimenting with regenerative techniques, although it still focuses on maintaining or increasing crop yields, reducing costs of crop production, and improving crop profits.[48]

In 2016 soil scientist John P. Reganold, a member of the National Research Council committee that oversaw preparation of the sustainable agriculture report, and his doctoral student Jonathan M. Wachter at Washington State University in Pullman published an important paper in which they measured the relative performance—past, present, and future—of organic agriculture in meeting the national academy goals. Based on their exhaustive review of published scientific studies, Reganold and Wachter reported that organic yields were lower (8 percent to 25 percent) and profits higher (29 percent to 32 percent) than yields from conventional farming, suggesting that the organic approach is good for producers as well as consumers. By paying premium prices for organic products, consumers get better quality and, in many cases, learn the sources of those products. Producers are compensated for their conservation efforts and for preserving rural landscapes, helping to keep alive the public's long-held sentimental attachment to rural life. In addition to its charm, organic farming generally requires more labor than does conventional farming, which means that more families contribute to the vitality of rural communities.[49] Some of us would like to believe that natural and organic agriculturists are the most committed to providing economic and societal benefits as well as to earning modest profits for themselves.

Arguably more personal and tangible than any other method of cultivation, organic agriculture teaches us to appreciate nature's capacities and limits

FIGURE 7.4. Restorative agriculture, Chatfield Farms, 2020. *Courtesy*, Denver Botanic Gardens Chatfield Farms.

and, by extension, the fact that we live in a finite world. As the ancient Greek philosophers expressed long ago, our civilization is based on the notion of infinite progress. Commitment to that notion was reaffirmed by our nation's founders and since then has been enhanced to mean unlimited growth and development. Given the urgency of adapting our cultural values to the realities of nature's limits, former governor Richard Lamm suggested facetiously that "we need not better scientists and technicians but better poets and philosophers." To be sure, scientists and technicians have performed wonders, extracting huge increases in yields from the earth, resulting in enough food to feed a growing global population. "Where past genius was recognized for pushing back [nature's] limits," Lamm explained, "future genius will be recognized on how to adapt to the very rapidly approaching limits [of nature]." That is where philosophers and poets come into play.[50]

More and more, Colorado agriculturists—organic, regenerative, conventional, and somewhere in between—are relying on technology in adapting to soil and water limits. Agricultural ethicists hold the view that technology can delay but not solve the dilemma of growth. Despite Lamm's expressed self-criticism that as governor he failed to rein in unfettered growth, one must hope that new generations of residents attracted to Colorado by its magnificent scenery have the moral courage to keep trying. Growing consumer demand for locally grown natural and organic food, combined with the popularity of open spaces and mountain landscapes, suggests some recognition that the much-vaunted Colorado lifestyle must be modified to be preserved. In his 1976 farm bureau interview, Lamm reminded readers that one can overgraze a field and also overgraze a state. Sustainable agriculture, natural and organic in particular, can show the way toward a new culture of limits paradigm.[51]

Epilogue

Variously described by early explorers as a "barren waste" and "uninhabitable by civilized man," the plains of eastern Colorado rank among the most difficult places in the United States to raise crops. Fewer than 60 miles separate the James Howell homestead near Flagler in western Kit Carson County and Kochis Farms in eastern Elbert County; the former was visited in 1900 by experiment station agent James E. Payne and the latter in 2019 by coauthor John Freeman. Both sites illustrate human ingenuity in adapting to the soil, climate extremes, and limited water. The Howells and their neighbors, according to Payne, had learned the hard way that they could not survive on dry-land farming, much less on their accustomed farming methods from further east. They had to create a self-contained farmstead to sustain themselves and their families. To begin with, Howell had purchased rootstock black locust, a non-native species known to develop rapidly into dense windbreaks; he had also dug a shallow well and a storm water reservoir for irrigating vegetables and small fruits and for watering livestock and planted orchard fruit seedlings, each in its own basin, to collect water.[1] He could not

https://doi.org/10.5876/9781646422050.c008

possibly have imagined the scientific advances and technical innovations that would enable Virgil and Jan Kochis and their son Michael to thrive on a moderately profitable dry-land grain/livestock operation while conserving the land. Since the 1980s the family has used minimal till, rotated winter wheat fields with other grains depending on market demand, and practiced holistic grazing, although they did not call it by that name. As a result of careful cultivation, the soil has become optimal for winter wheat; organic content is back to normal for the High Plains. Synthetic fertilizer is no longer required, and herbicides are used sparingly. To help offset grain and livestock market fluctuations, the Kochis family can depend on a new cash crop: royalties from wind turbines spread over the farm's 10,000 acres. Food on their table comes from the grocery store.

For more than a century, the Kochis family has benefited from the amenities of urban living as envisioned by President Theodore Roosevelt and implemented by President Franklin D. Roosevelt under the New Deal; urban dwellers have yearned for rural living and open spaces, whether cultivated fields, rangeland, or forests. No one better understands the overall cultural and health aspects of agriculture than Jan Kochis, a registered dietician and longtime chair of the Rocky Mountain Farmers Union. Even before that, beginning in the 1970s under farmers union auspices, Virgil and Jan Kochis worked to connect rural residents with urban audiences unfamiliar with the challenges of and opportunities in agriculture. Among farm organizations, the farmers union has the longest and most consistent record in emphasizing a holistic view of stewardship that consists not only of caring for the land and water but working to establish justice and equity and to put into practice the golden rule. From its beginning, the farmers union's special affinity for small to mid-scale farmers and ranchers extended as well to industrial workers and consumers, most recently to residents of Denver's low-income neighborhoods through urban agriculture.

As more consumers want to know how their food is grown and by whom, it is understandable that more people—mostly college-educated men and women with little or no prior experience—are seeking to pursue careers in farming and ranching. But the obstacles are daunting in trying to find land and secure capital. While technical assistance is readily available once land and capital are in place, only a few groups and agencies help with connecting land seekers and landowners. LandLink is one of those groups, part of a nationwide effort initiated by Iowa State University. LandLink's Colorado sponsor is Guidestone, a charitable nonprofit founded by organic farmer David Lynch of Buena Vista. His group's major challenge is finding farmers and ranchers willing to sell or lease all or part of their properties to beginning agriculturists because for a landowner, selling to developers is far more lucrative.[2] Another effort, not entirely successful, is sponsored by counties that lease both public and private agricultural lands under their management. In Boulder County, most lessees have been part-time and hobby farmers; county managers would prefer to lease 10-acre to 20-acre plots to full-time market gardeners engaged in high-value, high-intensity cultivation. A different effort entirely has been occurring on mostly rangeland on the Western Slope, especially around mountain communities, where county land-use regulations protect open spaces from unplanned and unfettered development. In Steamboat Springs and elsewhere, the extension service has assisted local nonprofit groups seeking ways to encourage small-scale natural and organic agriculture.

At present, natural and organic agriculture supplements conventional agriculture. Indeed, if agriculturists suddenly switched to all natural and organic, the world's population would find itself in trouble. But with the help of science and technology, the shift toward more sustainable agriculture is accelerating and with it a reassessment of what is meant by agricultural productivity. To be sure, sustainable yields of food and fiber remain the most measurable products of agriculture. In Colorado with its splendid scenery, land-use planners and social scientists are seeking to measure the value of agriculture's contributions to recreation and tourism, in particular agro-tourism, which is increasingly attractive as the population becomes more removed from its rural roots. Furthermore, there are the immeasurables such as the cultural value of agriculture in reminding us of rural life, preserving rural landscapes, and protecting scenic vistas.

Since time immemorial, the mark of a good farmer has been to leave the soil in better condition than it was when he or she found it. As Colorado's virgin soils became cultivated, losing some of their vitality, agriculturists and agricultural scientists focused more and more on ways to maintain and even improve the soil's capacity to supply crops with essential nutrients and to produce consistent yields. Today, the operative term is no longer soil fertility but soil health—a broader, more encompassing term that describes balancing the biological, chemical, and physical components of the soil, thereby preserving a balance among organisms in the soil and between those organisms and their general environment, the ecosystem. The indisputable fact that synthetic fertilizers, herbicides, and insecticides can disrupt those balances has caused most farmers, whether commodity growers or market gardeners, to adopt at least some aspects of natural and organic cultivation practices.

Arguably, a more strikingly obvious cause of imbalance is climate change. As soils warm up, conventional crops require more irrigation at a time when the supplies of water are diminishing. For agriculture heavily dependent on spring runoff, higher temperatures are altering the timing, making it earlier and more rapid. Neither agriculturists nor non-agriculturists need to be reminded that extreme weather patterns—drought, heat waves, wildfires—are becoming more frequent and more vehement. Here, too, adopting more natural, time-tested techniques such as adding organic matter and keeping soils covered at all times, combined with planting new crops such as perennial grains, represent cultivation practices to ensure healthy soil and mitigate water shortages. If the New Deal taught us anything about agriculture, it is that citizens through their government must provide the financial incentives for agriculturists to do what they should to protect our land and water. Put another way, agriculture is not just another way of buying and selling or making things. On behalf of all of us, farmers and ranchers exercise a sacred role as stewards of the earth.

As the ancient Greek philosophers expressed long ago, our civilization is based on a linear idea of unlimited and perpetual progress. Commitment to that idea was reaffirmed by our nation's founders. Given the ominous results of virtually unlimited growth and development—in agricultural terms, to override the limits of nature—we imperil more than land and water. Interestingly, organic agriculture presents us with an alternative circular or closed view of the idea of progress that recognizes the limits of nature. It was to reorient our understanding of progress that former governor Lamm suggested that we need more imaginative thinkers. Based on the extraordinary scientific discoveries and innovations of the past few decades, we know that it is technically possible to achieve that "balance between man and nature." As the eminent ecologist Paul B. Sears wrote during the heyday of the New Deal, "the real problem [of restoring balance] is moral."[3]

When Richard Lamm moved to Colorado in 1961, he believed it was still possible to prevent the haphazard, unrestrained development he had witnessed while attending law school in California and that through public leadership he could convince his fellow citizens to make minimal material sacrifices to preserve Colorado's healthy lifestyle. In particular, he was encouraged by the immense popularity of the National Western Stock Show, which he viewed as wistful affection for Colorado's agricultural past and hope for a future diversified economy that conserves the natural values that have attracted people to the state.

During the decades since Governor Lamm presented his framework for achieving a balance between people and nature, the state's population has doubled to 5.6 million and the economy has boomed accordingly, which led him to believe that the very future of life on earth depends on replacing our current "culture of growth" with a "culture of limits." Coming out of their own environmental activism in the 1970s, the French invented a melodious word to express that new paradigm: *decroissance* (opposite of *croissance*), translated as de-growth (opposite of growth), which has been defined as a reduction in production and consumption of goods that allows us to live "convivially and frugally." Sustainable de-growth requires that "future societies live within their ecological means."[4]

As we write this section, we find ourselves still masking and social distancing, grateful that scientists have provided us with the means to adapt to the presence of the Corona family of viruses. At the same time, we are experiencing more frequent and more extreme weather events, generally considered the result of two centuries of mostly unfettered economic development. If there is any comforting aspect of the pandemic and the disruptions—both social and economic—due to climate change, it compels us to reflect on what gives meaning to our lives. Will we be more willing than before to forego some material benefits so we can live more happily in a healthier environment? Publicist William Pabor's advice to pioneer farmers—adapt farming methods to the land and nature will take care of the rest—holds more meaning today than ever before.

Notes

Introduction

1. William E. Pabor, Colorado as an Agricultural State: Its Farms, Fields, and Garden Lands (New York: Orange Judd, 1883), 12 (quote), 65.

Chapter 1: Agricultural Foundations

1. *Greeley Tribune*, December 28, 1870, quoted in David Boyd, *A History: Greeley and the Union Colony of Colorado* (Greeley, CO: Greeley Tribune Press, 1890), 266.
2. David Boyd obituary, *Greeley Tribune*, December 12, 1908, 10.
3. Boyd, *A History*, 143.
4. Kelly Kindscher, *Edible Wild Plans of the Prairie: An Ethnobotanical Guide* (Lawrence: University Press of Kansas, 1987), 4, 79–83, 134–37, 146–48, 183–89.
5. Linda S. Cordell and Maxine E. McBrinn, *Archeology of the Southwest*, 3rd ed. (Walnut Creek, CA: Left Coast Press, 2012), 169; Brian M. Fagan, *Ancient North America. The Archeology of a Continent*, 3rd ed. (New York: Thames and Hudson, 2000), 334–35; Paul M. O'Rourke, *Frontier in Transition. A History of Southwestern Colorado* (Denver: US Bureau of Land Management, 1980), 15–18.

6. Cordell and McBrinn, *Archeology of the Southwest*, 129–33; Paul E. Minnis, ed., *People and Plants in Ancient Western North America* (Washington, DC: Smithsonian Books, 2004), 2025.

7. Carl Abbott, Stephen J. Leonard, and Thomas J. Noel, *Colorado: A History of the Centennial State*, 5th ed. (Boulder: University Press of Colorado, 2013), 35–36; Alvin T. Steinel, *History of Agriculture in Colorado* (Fort Collins, CO: State Agricultural College, 1926), 30.

8. Frederic J. Athearn, *Land of Contrast: A History of Southeast Colorado* (Denver: US Bureau of Land Management, 1985), 49–50; Gregory A. Hicks and Devon G. Peña, "Sin aqua no hay vida: Colorado's Acequias—a Water Democracy," in Karla A. Brown, ed., *Citizen's Guide to Colorado's Water Heritage* (Denver: Water Education Colorado, 2004), 15.

9. Steinel, *History of Agriculture*, 29–31, 177; Josiah Gregg, *Scenes and Incidents in the Western Prairies: During Eight Expeditions, and Including a Residence of Nearly Nine Years in Northern Mexico* (Philadelphia: J. W. Moore, 1857), 1: 149–51; Athearn, *Land of Contrast*, 52.

10. Horace Greeley, *An Overland Journey from New York to San Francisco in the Summer of 1859*, ed. Charles T. Duncan (New York: Alfred A. Knopf, 1964 [1860]), 127; Francis Parkman, *The Oregon Trail* (New York: Library of America, 1991 [1847]), 274.

11. Quoted in Leroy R. Hafen, ed., *Colorado and Its People* (New York: Lewis Historical Publishing, 1948), 1: 21; John F. Freeman, *High Plains Horticulture: A History* (Boulder: University Press of Colorado, 2008), 15–16.

12. William Byers quoted in Hafen, *Colorado and Its People* 1: 182–83; Steinel, *History of Agriculture*, 180–82.

13. Alfred Thomson [editor, *Miner's Register*, Central City] quoted in Steinel, *History of Agriculture*, 54.

14. *Rocky Mountain News*, August 1, 1860, quoted in Steinel, *History of Agriculture*, 186.

15. Steinel, *History of Agriculture*, 187.

16. Hafen, *Colorado and Its People* 1: 54–58.

17. Quoted in William E. Pabor, *Colorado as an Agricultural State: Its Farms, Fields, and Garden Lands* (New York: Orange Judd, 1883), 164–65.

18. John C. Frémont, *Report of the Exploring Expedition to the Rocky Mountains in the Year 1842, and to Oregon and North California in the Years 1843–44* (Washington, DC: Smithsonian Institution, 1988 [1845]), 114.

19. Steinel, *History of Agriculture*, 107–10, 451–53.

20. Steinel, *History of Agriculture*, 119.

21. James E. Payne, "Cattle Raising on the Plains," *Colorado Agricultural Experiment Station Bulletin* 87 (June 1904): 8.

22. Quoted in Lowell Baumunk, "Nathan Cook Meeker, Colonist," master's thesis, Colorado State College of Education, Greeley, 1949, 39–40; Freeman, *High Plains Horticulture*, 33–38.

23. Baumunk, "Meeker," 43; Certificate of Organization of the Union Colony of Colorado in James F. Willard, ed., *The Union Colony at Greeley, Colorado, 1869–1871* (Boulder: University of Colorado Historical Collections, 1918), 15–16; Robert G. Dunbar, "History of Agriculture," in Hafen, *Colorado and Its People* 2: 122–23.

24. Quoted in Steinel, *History of Agriculture*, 199.

25. Boyd, *A History*, 156; Steinel, *History of Agriculture*, 389–90 (quote, p. 390).

26. Boyd, *A History*, 154.

27. William Stuart, *The Potato: Its Culture, Uses, History, and Classification*, 4th ed. rev. (Chicago: J. B. Lippincott, 1937), 385–87.

28. Stuart, *The Potato*, 291; Boyd, *A History*, 153.

29. Boyd, *A History*, 154.

30. Boyd, *A History*, 144.

31. Philip Conford, *The Origins of the Organic Movement* (Edinburgh: Floris, 2001), 17.

32. Boyd, *A History*, 145; [Ainsworth E. Blunt?], "Alfalfa: Its Growth, Composition, Digestibility, etc.," *Colorado Agricultural Experiment Station Bulletin* 8 (July 1889): 3–4.

33. Boyd, *A Historyy*, 145–46 (quote, p. 145).

34. Boyd, *A History*, 152–53; Stuart, *The Potato*, 143.

35. P. Andrew Jones and Tom Cech, *Colorado Water Law for Non-Lawyers* (Boulder: University Press of Colorado, 2009), 60–68; Robert R. Crifasi, *A Land Made from Water: Appropriation and the Evolution of Colorado's Landscape, Ditches, and Water Institutions* (Boulder: University Press of Colorado, 2015), 155–82.

36. Quoted in Dunbar, "History of Agriculture," 126; see also Crifasi, *A Land Made from Water*, 155–82.

37. Freeman, *High Plains Horticulture*, 59; see also Robert G. Dunbar, *Forging New Rights in Western Water* (Lincoln: University of Nebraska Press, 1983), 88–98.

38. Crifasi, *A Land Made from Water*, 67–68; Steinel, *History of Agriculture*, 225–26.

39. Steinel, *History of Agriculture*, 202–3, 221.

40. Lowell K. Dyson, *Farmers' Organizations* (New York: Greenwood, 1986), 243.

41. David R. Berman, *Radicalism in the Mountain West, 1890–1920: Socialists, Populists, Miners, and Wobblies* (Boulder: University Press of Colorado, 2007), 42, 193, 199.

42. James E. Wright, *The Politics of Populism: Dissent in Colorado* (New Haven, CT: Yale University Press, 1974), 42–45.

43. History.com Editors, "Knights of Labor," https://www.history.com/topics/19th-century/knights-of-labor, accessed March 2020.

44. Robert C. McMath Jr., *American Populism: A Social History, 1877–1898* (New York: Hill and Wang, 1993), 115.

45. Berman, *Radicalism in the Mountain West*, 47.

46. James E. Wright, *The Politics of Populism*, 195.

47. Byers quoted in *Rocky Mountain News*, November 3, 1866, 1; Steinel, *History of Agriculture*, 282–86; Olivier de Serres, *Le Théâtre d'Agriculture et Mesnage des Champs* (Arles, France: Actes Sud, 2001 [1599]), 824.

48. William E. Pabor, *First Annual Report of the Union Colony of Colorado* (New York: George W. Southwick, 1871), 17; William J. May, *The Great Western Sugarlands: The History of the Great Western Sugar Company and the Economic Development of the Great Plains* (New York: Garland, 1989), 27–29.

49. Hafen, *Colorado and Its People* 1: 512.

50. An Act Donating Public Lands to the Several States and Territories Which May Provide Colleges for the Benefit of Agriculture and the Mechanic Arts, *US Statutes at Large* 12 (1862): 504.

51. James E. Hansen III, *Democracy's College in the Centennial State: A History of Colorado State University* (Fort Collins: Colorado State University, 1977), 23, 28, 41, 48, 51.

52. Hansen, *Democracy's College*, 29–31, 47; Ainsworth E. Blount, "Report of Experiments in the Farm Department on Grains, Grasses, and Vegetables," *Colorado Agricultural Experiment Station Bulletin* 2 (December 1887): 2–10.

53. James R. Kluger, *Turning on Water with a Shovel: The Career of Elwood Mead* (Albuquerque: University of New Mexico Press, 1992), 7–13; Elwood Mead, "Report on Experiments in Irrigation and Meteorology," *Colorado Agricultural Experiment Station Bulletin* 1 (August 1887): 3–5; Hansen, *Democracy's College*, 70–72.

54. Hansen, *Democracy's College*, 73.

55. James Cassidy, "Report of Experiments with Potatoes and Tobacco," *Colorado Agricultural Experiment Station Bulletin* 4 (February 1888): 12–13 (quote, p. 12).

56. James Cassidy, "Notes on Insects and Insecticides 1888," *Colorado Agricultural Experiment Station Bulletin* 6 (January 1889): 2–4.

57. An Act to Establish Agricultural Experiment Stations, *US Statutes at Large* 24 (1887): 440–42.

58. James Cassidy and David O'Brine, "Potatoes and Sugar Beets," *Colorado Agricultural Experiment Station Bulletin* 7 (April 1889): 21–23; Frank L. Watrous, "Sugar Beets; Irish Potatoes; Fruit-Raising," *Colorado Agricultural Experiment Station Bulletin* 21 (October 1892): 3.

59. May, *Great Western Sugarlands*, 3.

60. May, *Great Western Sugarlands*, 4–7.

61. Steven F. Mehls, *The Valley of Opportunity: A History of West-Central Colorado* (Denver: US Bureau of Land Management, 1982), 142–44; "History of Weld County's Farm Workers," *Greeley Tribune*, November 2002, https://www.greeley

tribune.com/news/local/the-history-of-welds-farm-workers/, accessed March 2020.

62. Mehls, *Valley of Opportunity*, 131, 138–39.

63. Pabor, *Colorado as an Agricultural State*, 167.

64. O'Rourke, *Frontier in Transition*, 130.

65. G. Michael McCarthy, "White River Forest Reserve: The Conservation Conflict," *Colorado Magazine* 49 (Winter 1972): 56, 66.

66. Freeman, *High Plains Horticulture*, 108.

67. 1870 and 1900 US Census. See bibliographical note, page 257.

68. Pabor, *Colorado as an Agricultural State*, 195 (first quote), 201 (second quote).

Chapter 2: Making the Land Flourish

1. Alvin T. Steinel, *History of Agriculture in Colorado* (Fort Collins, CO: State Agricultural College, 1926), 277–78; J. W. Adams, "Adobe as a Building Material for the Plains," *Colorado Agricultural Experiment Station Bulletin* 174 (1910): 1–8; KSU Alumni Association, *Record of the Alumni of the Kansas State Agricultural College* (Manhattan: Kansas State University, 1914), 108, https://archive.org/stream/recordofalumniofookansrich#page/108/mode/2up, accessed April 2016.

2. Steinel, *History of Agriculture*, 255, 279–80 (quote).

3. Theodore Roosevelt, "First Annual Message," December 3, 1901, in Gerhard Peters and John T. Woolley, *The American Presidency Project*, https://www.presidency.ucsb.edu, accessed April 2019; G. Michael McCarthy, *Hour of Trial: The Conservation Conflict in Colorado and the West, 1891–1907* (Norman: University of Oklahoma Press, 1977), 66–67, 159–60.

4. An Act Appropriating the Receipts from the Sale and Disposal of Public Lands in Certain States and Territories to the Construction of Irrigation Works for the Reclamation of Arid Lands, *US Statutes at Large* 32 (1902): 388–90.

5. Steinel, *History of Agriculture*, 527–39; Shelly C. Dudley, "The First Five: A Brief History of the Uncompahgre Project (Gunnison)," http://www.waterhistory.org/histories/reclamation/uncompahgre/, accessed April 2016; US Census 1920.

6. Liberty Hyde Bailey, ed., *Report of the Commission on Country Life* (New York: Sturgis and Walton, 1911), 24 (first quote), 84 (second and third quotes); see Scott J. Peters and Paul A. Morgan, "The Country Life Commission: Reconsidering a Milestone in American Agricultural History," *Agricultural History* 78, no. 3 (Summer 2004): 289–316.

7. Bailey, *Report of the Commission on Country Life*, 83–85, 89.

8. Liberty Hyde Bailey, *The Holy Earth: Toward a New Environmental Ethic*, intro. Norman Wirzba (Mineola, NY: Dover, 2009), 16 (first quote), 53 (second quote); Zachary Michael Jack, ed., *Liberty Hyde Bailey: Essential Agrarian and Environmental Writings* (Ithaca, NY: Cornell University Press, 2008), 6, 28–30.

9. Bailey, *Holy Earth*, 9 (first quote), 24 (second quote).

10. US Census 1900 and 1920.

11. Jethro Tull, *Horse-Hoeing Husbandry: or, an Essay on the Principles of Vegetation and Tillage* (London: A. Millar, 1762); John F. Freeman, *High Plains Horticulture: A History* (Boulder: University Press of Colorado, 2008), 52–53, 118.

12. James E. Payne, "Unirrigated Lands of Eastern Colorado," *Colorado Agricultural Experiment Station Bulletin* 77 (February 1903): 15, and "Wheat Raising on the Plains," *Colorado Agricultural Experiment Station Bulletin* 89 (June 1904): 25–27.

13. James E. Payne, "Unirrigated Alfalfa on Upland," *Colorado Agricultural Experiment Station Bulletin* 90 (June 1904): 31–33, and "Field Notes from Trips in Eastern Colorado," *Colorado Agricultural Experiment Station Bulletin* 59 (December 1900): 12–13.

14. Payne, "Field Notes," 16.

15. James E. Payne, "Cattle Raising on the Plains," *Colorado Agricultural Experiment Station Bulletin* 87 (June 1904): 5, 9 (quote).

16. "Philo K. Blinn," *Rocky Mountain Collegian* 3, no. 9 (May–June 1894): 66, and *Collegian* 18, no. 11 (March 1909), 16; James E. Hansen III, *Democracy's College in the Centennial State: A History of Colorado State University* (Fort Collins: Colorado State University, 1977), 212; Philo K. Blinn, "Early Cantaloupes," *Colorado Agricultural Experiment Station Bulletin* 95 (December 1904): 3, and "Development of the Rocky Ford Cantaloupe Industry," *Colorado Agricultural Experiment Station Bulletin* 108 (March 1906): 3, and "A Rust-Resistant Cantaloupe," *Colorado Agricultural Experiment Station Bulletin* 104 (November 1905): 5.

17. Philo K. Blinn, "Cantaloupe Breeding," *Colorado Agricultural Experiment Station Bulletin* 126 (January 1908): 9–11.

18. Blinn, "Early Cantaloupes," 4; see Wendell Paddock, "Tillage, Fertilizers, and Shade Crops for Orchards," *Colorado Agricultural Experiment Station Bulletin* 142 (May 1909): 5–9; Robert G. Dunbar, "History of Agriculture," in Leroy Hafen, ed., *Colorado and Its People* (New York: Lewis Historical Publishing, 1948), 2: 144.

19. Hansen, *Democracy's College*, 160–62; Wayne D. Rasmussen, *Taking the University to the People: Seventy-five Years of Cooperative Extension* (Ames: Iowa State University Press, 1989), 31.

20. Joseph Cannon Bailey, *Seaman A. Knapp: Schoolmaster of American Agriculture* (New York: Columbia University Press, 1945), 154–57; Freeman, *High Plains Horticulture*, 64, 173.

21. Hansen, *Democracy's College*, 188; L. M. Winsor, San Luis Valley, "1912–13 Annual Report," box 17, folder 18, 1, Records of the Colorado Cooperative Extension, Agricultural and Natural Resources Archive, Colorado State University, Fort Collins (hereafter Records of CES).

22. George H. Glover, "Hog Cholera Control," *Colorado Agricultural Experiment Station Bulletin* 197 (May 1914): 3–4, 6–7.

23. Winsor, "1912–13 Annual Report," 36–38, Records of CES; William P. Headden, "How Can We Maintain the Fertility of Our Colorado Soils?" *Colorado Agricultural Experiment Station Bulletin* 99 (March 1905): 3, 6–7.

24. Winsor, "1912–13 Annual Report," 42–43, Records of CES.

25. Dunbar, "History of Agriculture," 2: 156; James E. Hansen III, *Beyond the Ivory Tower: A History of Colorado State University Cooperative Extension* (Fort Collins: Colorado State University, 1991), 20–22.

26. Quoted in Bailey, *Report of the Commission on Country Life*, 42–43.

27. An Act to Provide for Cooperative Agricultural Extension Work between the Agricultural Colleges, *US Statutes at Large* 38 (1914): 372 (first quote); Bailey, *Report of the Commission on Country Life*, 43 (second quote).

28. Dexter W. Lillie, Cheyenne County, "Annual Report [1941]," box 16, folder 41, preface (n.p.), 6, Records of CES; Alvin B. Kezer, "Forage Crops for the Colorado Plains," *Colorado Agricultural Experiment Station Bulletin* 214 (October 1915): 5; Hansen, *Beyond the Ivory Tower*, 32; David B. Danbom, *Born in the Country: A History of Rural America*, 2nd ed. (Baltimore: Johns Hopkins University Press, 2006), 176–77.

29. Samuel C. Salmon, "Establishing Kanred Wheat in Kansas," *Kansas Agricultural Experiment Station Circular* 74 (August 1919): 3–4; J. Allen Clark and Samuel C. Salmon, "Kanred Wheat," *USDA Circular* 194 (September 1941): 3, 8.

30. A. E. McClymonds, "Annual Report [1920]," box 109, folder 25, 1, 54–58, Records of CES.

31. Steinel, *History of Agriculture in Colorado*, 442–43.

32. Bruce J. Thornton, "The Colorado Pure Seed Law and Use of the Seed Laboratory," *Colorado Agricultural Experiment Station Miscellaneous Series* 129 (1942?), 1–2; Freeman, *High Plains Horticulture*, 178–79.

33. A. E. McClymonds, "Annual Report [1920]," box 109, folder 25, 2, Records of CES; J. E. Morrison quoted in R. P. Yates, "Historical Appraisal of the Seed Registration Phase of the Crops Project," Annual Report of Assistant Extension Agronomist (1939), box 110, folder 17, 2, Records of CES.

34. Yates, "Historical Appraisal," 1–3 (quote on page 2).

35. Wilfred W. Robbins and Breeze Boyack, "The Identification and Control of Colorado Weeds," *Colorado Agricultural Experiment Station Bulletin* 251 (July 1919): 6,

21–26, 33–34; Robert L. Zimdahl, *Agriculture's Ethical Horizon*, 2nd ed. (Amsterdam: Elsevier, 2012), 30.

36. Robbins and Boyack, "Identification and Control of Colorado Weeds," 10–11.

37. James E. Payne, "Unirrigated Alfalfa on Upland," *Colorado Agricultural Experiment Station Bulletin* 90 (June 1904): 33; Charles R. Jones, "Grasshopper Control," *Colorado Agricultural Experiment Station Bulletin* 233 (June 1917): 7–9.

38. John F. Freeman, *Persistent Progressives: The Rocky Mountain Farmers Union* (Boulder: University Press of Colorado, 2016), 6–7.

39. Freeman, *Persistent Progressives*, 17, 22.

40. *Peetz* (CO) *Gazette*, November 7, 1919.

41. L. V. "Lew" Toyne, "Early History of the Farm Bureau in Colorado," 1969, typescript, 1–3, box 1, folder 2, Colorado Farm Bureau Collection, Agricultural and Natural Resources Archive, Colorado State University, Fort Collins (quote); Nell Brown Propst, *Grassroots People* (Denver: Colorado Farm Bureau, 1997), 4.

42. Michael J. Lansing, *Insurgent Democracy: The Nonpartisan League in North American Politics* (Chicago: University of Chicago Press, 2015), x.

43. Lowell K. Dyson, *Farmers' Organizations* (New York: Greenwood, 1986), 257–69.

44. Lansing, *Insurgent Democracy*, 85; Freeman, *Persistent Progressives*, 44; George L. Bickel, *Rocky Mountain Farmers Union: A History 1907–1978* (Denver: Rocky Mountain Farmers Union, 1978), 6.

45. Lansing, *Insurgent Democracy*, 142, 183, 221.

Chapter 3: Stretching Nature's Limits

1. Thomas H. Summers and Warren H. Leonard, "The Story of a Successful Farm in Northern Colorado," *Extension Services Circular* 48 (October 1926): 1–4.

2. Emil P. Sandsten, "Potato Growing in Colorado," *Colorado Agricultural Experiment Station Bulletin* 314 (January 1927): 3; Theodore Saloutos, *The American Farmer and the New Deal* (Ames: Iowa State University Press, 1982), 5–14.

3. Frank R. Lamb, Conejos County, "Annual Report [1925]," box 17, folder 22, 2, Records of the Colorado Cooperative Extension, Agricultural, and Natural Resources Archive, Colorado State University, Fort Collins (hereafter Records of CES).

4. Waldo Kidder, "Improved and Registered Seed for Colorado," included in "Annual Report [1922] of the Extension Agronomist," box 109, folder 27, 1 Records of CES.

5. Kidder, "Improved and Registered Seed for Colorado," 1–3.

6. Waldo Kidder, "Annual Report [1925] of the Extension Agronomist," box 110, folder 1, 32, Records of CES; J. C. Hackelman and W. O. Scott, *A History of Seed Certification in the United States and Canada* (Raleigh, NC: Association of Official Seed Certification Agencies, 1990), 3.

7. William M. Case, "Annual Report [1931] of the Acting Extension Horticulturist," box 120, folder 11, 6–7; T. G. Stewart, "Annual Report [1931] of the Extension Agronomist," box 110, folder 6, 63; R. P. Yates, "Annual Report [1939] of the Assistant Extension Agronomist," box 110, folder 17, all in Records of CES.

8. Howard S. MacMillan quoted in *Colorado Potato Grower* 4, no. 4 (October 1926): 3, in box 9, folder 16, Records of the Colorado Potato Growers Exchange, Agricultural and Natural Resources Archive, Colorado State University, Fort Collins (hereafter Records of CPG).

9. Sandsten, "Potato Growing in Colorado," 5.

10. W. A. Henry, "Professor Sandsten Called to Wisconsin," *Wisconsin Horticulturist* 7, no. 5 (July 1902): 8; James E. Hansen III, *Democracy's College in the Centennial State: A History of Colorado State University* (Fort Collins: Colorado State University, 1977), 216.

11. US Census 1920 and 1930; Albert A. Goodman, Rio Grande County, "Annual Report [1930]," box 74, folder 31, 34, Records of CES.

12. Sandsten, "Potato Growing in Colorado," 5–6.

13. Alvin T. Steinel, *History of Agriculture in Colorado* (Fort Collins, CO: State Agricultural College, 1926), 306.

14. William J. May, *The Great Western Sugarlands: The History of the Great Western Sugar Company and the Economic Development of the Great Plains* (New York: Garland, 1989), 159–60, 167–68, 174.

15. Waldo Kidder, "Annual Report [1927]," box 10, folder 2, 91, Records of CES.

16. May, *Great Western Sugarlands* 153, 159–60.

17. John E. Dalton, *Sugar: A Case Study of Government Control* (New York: MacMillan, 1937), 146–50.

18. John F. Freeman, *Persistent Progressives: The Rocky Mountain Farmers Union* (Boulder: University Press of Colorado, 2016), 26.

19. Freeman, *Persistent Progressives*, 42; Steinel, *History of Agriculture*, 323–24, 338–39.

20. J. C. Hale, Routt County, "Annual Report [1920]," box 76, folder 9, 8, Records of CES; Steinel, *History of Agriculture*, 575.

21. William L. Burnett, "Some Colorado Farm Pests with Suggestions for Control," *Colorado Agricultural Experiment Station Circular* 35 (June 1922): 4.

22. Floyd D. Moon, Routt County, "Annual Report [1930]," box 76, folder 12, 19, Records of CES.

23. George R. Smith, Boulder County, "Annual Report [1927]," box 11, folder 22, 10–11, Records of CES.

24. T. G. Stewart, "Annual Report [1931]," box 110, folder 6, 73, Records of CES.

25. R. H. Tucker, Delta County, "Annual Report [1931]," box 19, folder 62, 14–15, Records of CES.

26. Vaclav Smil, *Enriching the Earth: Fritz Haber, Carl Bosch, and the Transformation of World Food Production* (Cambridge, MA: MIT Press, 2001), 56.

27. Smil, *Enriching the Earth*, 61–107.

28. US Census 1930; "Phosphate Needed on Colorado Farms" and "Phosphates Can Be Secured by Members," *Colorado Potato Grower* 7, no. 8 (February 1930): 4, box 9, folder 16, Records of CPG.

29. William M. Case and T. G. Stewart, "A Word of Warning," inserted in Case, "Annual Report [1931]," box 120, folder 11, Records of CES, and "Annual Report [1931]," 7–8.

30. Golden Jubilee Pageant transcript, box 2, folder 47, 9, Records of the Rocky Mountain Farmers Union, Agricultural and Natural Resources Archive, Colorado State University, Fort Collins.

31. Kimberly K. Porter, "Embracing the Pluralistic Perspective: The Iowa Farm Bureau and the McNary-Haugen Movement," *Agricultural History* 74, no. 2 (Spring 2000): 382–83; Lee Egerstrom, Milton Hakel, and Bob Denman, *Connecting America's Farmers with America's Future: The National Farmers Union, 1902–2002* (Red Wing, MN: Lone Oak Press, 2002), 57.

32. *Colorado Union Farmer*, March 11, 1927, quoted in Freeman, *Persistent Progressives*, 43.

33. Edwin L. Earp, "The Rural Church," *Dickensonian* 45, no. 21 (March 21, 1918): 10–11, http://archives.dickinson.edu/dickinsonian/dickinsonian-march-21-1918, accessed March 2019.

34. Edwin L. Earp, *Biblical Backgrounds for the Rural Message* (New York: Association Press, 1922), 5–6, https://babel.hathitrust.org/cgi/pt?id=coo.31924029336926;view=1up;seq=7, accessed March 2019.

35. Earp, *Biblical Backgrounds*, 20.

36. Agricultural Marketing Act, *US Statutes at Large* 46 (1929): 11–19.

37. C. Fred Williams, "William M. Jardine and the Foundations for Republican Farm Policy, 1925–1929," *Agricultural History* 70, no. 2 (Spring 1996): 218–20, 226–29.

38. Albert A. Goodman, "Annual Report [1932]," box 74, folder 36, 90, 130–31, Records of CES.

39. James Wickens, *Colorado in the Great Depression* (New York, Garland, 1979), 9–10; William M. Case, "Annual Report [1931]," box 120, folder 11, Records of CES.

40. Wickens, *Colorado in the Great Depression*, 219–20; Steven F. Mehls, *The New Empire of the Rockies: A History of Northeast Colorado* (Denver: US Bureau of Land Management, 1984), 156.

41. Leroy Hafen, ed., *Colorado and Its People* (New York: Lewis Historical Publishing, 1948), 1: 550; Lowell K. Dyson, *Farmers' Organizations* (New York: Greenwood, 1986), 85; John L. Shover, *Cornbelt Rebellion: The Farmers' Holiday Association* (Urbana: University of Illinois Press, 1965), 213.

42. Case, "Annual Report [1931]," 2 (quote); Hansen, *Democracy's College*, 297.

43. Wickens, *Colorado in the Great Depression*, 12–14.

44. Wickens, *Colorado in the Great Depression*, 31–32; Emergency Relief and Construction Act, *US Statutes at Large* 47 (1932): 709–24.

45. Wayne D. Rasmussen, Gladys L. Baker, and James S. Ward, "A Short History of Agricultural Adjustment, 1933–75," *USDA Economic Research Service Agriculture Information Bulletin* 391 (March 1976): 2.

Chapter 4: Federal Engagement in Agriculture

1. Fred C. Case, Baca County, "Annual Report [1935]," box 8, folder 46, Records of the Colorado Cooperative Extension, Agricultural and Natural Resources Archive, Colorado State University, Fort Collins (hereafter Records of CES).

2. Anonymous letter attached to Case, "Annual Report [1935]"; see Douglas Sheflin, "The New Deal Personified: A. J. Hamman and the Cooperative Extension Service in Colorado," *Agricultural History* 90, no. 3 (Summer 2016): 361.

3. James Wickens, *Colorado in the Great Depression* (New York: Garland, 1979), 230–31; Glen R. "Spike" Ausmus, discussion with John Freeman, Springfield, CO, October 2018.

4. Quoted in Edward K. Muller, ed., *DeVoto's West: History, Conservation, and the Public Good* (Athens, OH: Swallow, 2005), 88.

5. Wickens, *Colorado in the Great Depression*, 57–58, 128–30, 137–43.

6. Wickens, *Colorado in the Great Depression*, 59–63; memorandum from Assistant Extension Director James E. Morrison to staff, March 3, 1933, included with William M. Case, "Annual Report [1933]," box 120, folder 13, Records of CES.

7. An Act to Relieve the Existing National Economic Emergency by Increasing Agricultural Purchasing Power, to Raise Revenue for Extraordinary Expenses Incurred by Reason of Such Emergency . . . *US Statutes at Large* 48, part 1 (1933): 32.

8. Act to Relieve the Existing National Economic Emergency, 31–54; Wayne D. Rasmussen, Gladys L. Baker, and James S. Ward, "A Short History of Agricultural Adjustment, 1933–75," *USDA Economic Research Service Agriculture Information Bulletin* 391 (March 1976): 2.

9. Albert A. Goodman, Rio Grande County, "Annual Report [1934]," box 74, folder 40, 112, Records of CES; James E. Hansen III, *Democracy's College in the Centennial State: A History of Colorado State University* (Fort Collins: Colorado State University, 1977), 297.

10. T. G. Stewart, "Annual Report [1934]," box 110, folder 9, 2, Records of CES; Wickens, *Colorado in the Great Depression*, 224–25; Robert G. Dunbar, "History of Agriculture," in Leroy Hafen, ed., *Colorado and Its People* (New York: Lewis Historical Publishing, 1948), 2: 135–36.

11. Dunbar, "History of Agriculture," 2: 223; David B. Danbom, *Born in the Country: A History of Rural America*, 2nd ed. (Baltimore: Johns Hopkins University Press, 2006), 211–12.

12. Fred Greenbaum, *Fighting Progressive: A Biography of Edward P. Costigan* (Washington, DC: Public Affairs Press, 1971), 143–54; Wickens, *Colorado in the Great Depression*, 235–37.

13. Raymond T. Burdick, "Economics of Sugar Beet Production in Colorado," *Colorado Agricultural Experiment Station Bulletin* 453 (June 1939): 47; Lorena Hickok to Harry L. Hopkins, en route Denver to Los Angeles, June 13, 1934, in Richard Lowitt and Maurine Beasley, eds., *One-Third of a Nation: Lorena Hickok Reports on the Great Depression* (Urbana: University of Illinois Press, 1981), 287–89; Wickens, *Colorado in the Great Depression*, 100, 238.

14. Wickens, *Colorado in the Great Depression*, 94–98, 265.

15. Wickens, *Colorado in the Great Depression*, 242–43.

16. Edson W. Barr, Routt County, "Annual Report [1940]," box 77, folder 2, 1, Records of CES; Dunbar, "History of Agriculture," 2: 156; see also Gene Wunderlich, *American Country Life: A Legacy* (Lanham, MD: University Press of America, 2003), 80–81.

17. Herbert G. Hansen, "Re-vegetation of Waste Range Land," *Colorado Agricultural Experiment Station Bulletin* 332 (January 1928): 4, 7; Thomas H. Summers and R. W. Schafer, eds., *An Agricultural Program for Northwest Colorado* (Fort Collins: Colorado Agricultural College Extension Service, 1928), 9.

18. An Act to Stop Injury to the Public Grazing Lands by Preventing Overgrazing and Soil Deterioration, to Provide for Their Orderly Use, Improvement, and Development, to Stabilize the Livestock Industry Dependent upon the Public Range, and for Other Purposes, *US Statutes at Large* 48, part 1 (1934): 1269–72; Steven C. Schulte, *Wayne Aspinall and the Shaping of the American West* (Boulder: University Press of Colorado, 2002), 22–23.

19. Wickens, *Colorado in the Great Depression*, 139–42.

20. An Act to Encourage National Industrial Recovery, to Foster Fair Competition, and to Provide for the Construction of Certain Useful Public Works, and for

Other Purposes, *US Statutes at Large* 48, part 1 (1933): 201; Gladys L. Baker, Wayne D. Rasmussen, Vivian Wiser, and Jane M. Porter, *Century of Service: The First 100 Years of the United States Department of Agriculture* (Washington, DC: US Department of Agriculture, 1963), 138–39, 191; Hugh H. Bennett and William R. Chapline, "Soil Erosion: A National Menace," *USDA Circular* 33 (1928): 2, 22.

21. An Act to Provide for the Protection of Land Resources against Soil Erosion, and for Other Purposes, *US Statutes at Large* 49, part 1 (1935): 163–64; "Eighty Years Helping People Help the Land: A Brief History of NRCS," http://www.nrcs.usda.gov/wps/portal/nrcs/detail/national/about/history/?cid=nrcs143_021392, accessed October 2016.

22. Wickens, *Colorado and the Great Depression*, 260–61.

23. Franklin D. Roosevelt, "Statement on Signing the Soil Conservation and Domestic Allotment Act," March 1, 1936, in Gerhard Peters and John T. Wooley, eds., *The American Presidency Project*, ucsb.edu, accessed October 2016.

24. An Act to Promote the Conservation and Profitable Use of Agricultural Land Resources by Temporary Federal Aid to Farmers and by Providing for a Permanent Policy of Federal Aid to States for Such Purposes, *US Statutes at Large* 49, part 1 (1936): 1148.

25. Douglas Helms, "The Preparation of the Standard State Soil Conservation Districts Law: An Interview with Philip M. Glick" [occasional paper], USDA Soil Conservation Service (February 1990): ii, http://www.nrcs.usda.gov/Internet/FSE_DOCUMENTS/nrcs143_021270.pdf, accessed October 2018.

26. Colorado Soil Conservation District Act, May 6, 1937, *Colorado Revised Statutes* 1953, vol. 5, chapter 28, 1, box 2, folder 3, Records of the Colorado Association of Soil Conservation Districts, Water Resources Archive, Colorado State University, Fort Collins (hereafter cited as CASCD); Dunbar, "History of Agriculture," 2: 156.

27. R. Neil Samson, *For Love of the Land: A History of the National Association of Conservation Districts* (League City, TX: National Association of Conservation Districts, 1985), 22–23, 39; Sheflin, "The New Deal Personified," 357–58.

28. Albert A. Goodman, "Annual Report [1936]," box 74, folder 43, 95, and "Annual Report [1937]," box 74, folder 44, 4–5; C. M. Knight, "Annual Report [1939]," box 17, folder 31, 13, all in Records of CES.

29. C. L. Hart, Cheyenne County, "Annual Report [1936]," box 16, folder 36, 2, Records of CES; Wickens, *Colorado and the Great Depression*, 261.

30. Joseph F. Brandon and Alvin Kezer, "Soil Blowing and Its Control in Colorado," *Colorado Agricultural Experiment Station Bulletin* 419 (January 1936): 14.

31. J. G. Lindley, "Protecting Colorado's Range Lands," *Soil Conservation* 1, no. 9 (April 1936): 12–13.

32. Dorothy C. Hogner, *Westward, High, Low, and Dry* (New York: E. P. Dutton, 1938), 295–302 (quote, p. 295); see also C. L. Hart, "Annual Report [1937]," box 16, folder 37, 3–4, Records of CES.

33. Raymond H. Skitt, Baca County, "Annual Report [1938]," box 8, folder 48, 1, 61, Records of CES.

34. Skitt, "Annual Report," 73; Sheflin, "The New Deal Personified," 367–68.

35. Ralph L. Parshall, "The Parshall Measuring Flume," *Colorado Agricultural Experiment Station Bulletin* 423 (March 1936): 3; Hansen, *Democracy's College in the Centennial State*, 226–27.

36. Daniel Tyler, *The Last Water Hole in the West: The Colorado–Big Thompson Project and the Northern Colorado Water Conservancy District* (Niwot: University Press of Colorado, 1992), 46.

37. Tyler, *Last Water Hole*, 41.

38. Tyler, *Last Water Hole*, 45.

39. Tyler, *Last Water Hole*, 58–80.

40. Barbara Mitchell, "Dr. Sudan," in *Grand County Historical Association Voices of the Past*, n.d., https://stories.grandcountyhistory.org/category/health-care, accessed October 2016.

41. Louis G. Davis, Jefferson County, "Annual Report [1935]," box 37, folder 5, 37, Records of CES.

42. Rasmussen, "Short History of Agricultural Adjustment," 6–8; Wickens, *Colorado in the Great Depression*, 271–74.

43. John F. Freeman, *Persistent Progressives: The Rocky Mountain Farmers Union* (Boulder: University Press of Colorado, 2016), 12, 71.

44. William M. Case, "Developing a Marketing Agreement Which Commands Popular Support," *Extension Service Review*, October 1939, included in Case, "Annual Report [1939]," box 120, folder 18, 47, Records of CES.

45. San Luis Valley Potato Improvement Association, "The New McClure Potato," 26, filed with W. Frank McGee [assistant horticulturist], "Annual Report [1941]," box 120, folder 14, Records of CES.

46. US Census 1930 and 1940.

47. K. A. McCaskill, Jefferson County, "Annual Report [1920]," box 37, folder 1, 1–3, Records of CES.

48. Jeanne Weaver, Jefferson County, "Annual Report [1937]," box 37, folder 6, 1; Louis G. Davis, "Annual Report [1938]," box 37, folder 8, preface, both in Records of CES; US Census 1940.

49. Louis G. Davis, "Annual Report [1935]," box 37, folder 4, 47 (quote), and "Annual Report [1934]," box 37, folder 3, 12–13; Charles M. Drage, Jefferson County,

"Annual Report [1938]," box 37, folder 8, 40, "Annual Report [1939]," box 37, folder 10, 2, and "Annual Report [1941]," box 37, folder 14, 54, all in Records of CES.

50. Charles M. Drage, "Annual Report [1938]," box 37, folder 8, 2, "Annual Report [1939]," box 37, folder 10, 1, 7, and "Annual Report [1940]," box 37, folder 12, 2, all in Records of CES.

51. Louis G. Davis, "Annual Report [1937]," box 37, folder 6, 15; Charles M. Drage, "Annual Report [1941]," box 37, folder 14, 32, and "Annual Report [1942]," box 37, 15, 36, all in Records of CES.

52. Don F. Hadwiger, "Henry A. Wallace, Champion of a Durable Agriculture," *American Journal of Alternative Agriculture* 8, no. 1 (1993): 3; Kevin M. Lowe, *Baptized with the Soil: Christian Agrarians and the Crusade for Rural America* (New York: Oxford, 2016), 83, 85, 162; "Soil Stewardship Committee Report," minute book, twenty-first annual meeting, Denver, January 3, 1966, 94–99, Records of CASCD; Michael J. Woods, *Cultivating the Soil: Twentieth-Century Catholic Agrarians Embrace the Liturgical Movement* (Collegeville, MN: Liturgical Press, 2009), 11, 20, 120.

53. Baker et al., *Century of Service*, 283, 305; Wayne D. Rasmussen, *Taking the University to the People: Seventy-five Years of Cooperative Extension* (Ames: Iowa State University Press, 1989), 107–12.

54. Danbom, *Born in the Country*, 230; Hansen, *Democracy's College in the Centennial State*, 35; V. D. Bailey, Conejos County, "Annual Report [1943]," 49–50, box 17, folder 35; Albert A. Goodman, "Annual Report [1945]," 35, box 75, folder 7, both in Records of CES.

55. Rasmussen, *Taking the University to the People*, 110; Charles M. Drage, "Annual Report [1942]," box 37, folder 15, 21, Records of CES; James E. Hansen III, *Beyond the Ivory Tower: A History of Colorado State University Cooperative Extension* (Fort Collins: Colorado State University, 1991), 56.

56. Robert (Bob) and Robert (R. T.) Sakata, discussion with John Freeman, Brighton, CO, September 2018; https://amache.org/agriculture-at-amache.org, accessed December 2018; Bill Hosokawa, *Colorado's Japanese Americans from 1886 to the Present* (Boulder: University Press of Colorado, 2005), 103–4; Frederic J. Athearn, *Land of Contrast: A History of Southeast Colorado* (Denver: US Bureau of Land Management, 1985), 172.

57. Charles M. Drage, "Annual Report [1945]," box 120, folder 17, 70, 75, 101–2; Drage and George M. List, "Suggestions Regarding the Use of DDT by Civilians," *Colorado Extension Services Circular* 1991 (November 1945), both in Records of CES.

58. Albert A. Goodman, "Annual Report [1945]," box 75, folder 7, 14, Records of CES.

59. Freeman, *Persistent Progressives*, 99–101.

Chapter 5: Advances in Productivity

1. Carl H. Powell, Delta County, "Program of Work [1959]," box 21. folder 4, 3–6, and "Annual Report [1960]," box 21, folder 5, 2, both in Records of the Colorado Cooperative Extension, Agricultural and Natural Resources Archive, Colorado State University, Fort Collins (hereafter Records of CES).

2. US Census 1945, 1964, and 1974.

3. Gladys L. Baker, Wayne D. Rasmussen, Vivian Wiser, and Jane M. Porter, *Century of Service: The First 100 Years of the United States Department of Agriculture* (Washington, DC: US Department of Agriculture, 1963), 364.

4. Douglas Helms, "Brief History of the USDA Soil Bank Program," *USDA Historical Insights* 1 (January 1985): 1; Wayne D. Rasmussen, Gladys L. Baker, and James S. Ward, "A Short History of Agricultural Adjustment, 1933–75," *USDA Economic Research Service Agriculture Information Bulletin* 391 (March 1976): 12–13.

5. Douglas Helms, "Great Plains Conservation Program, 1956–1981," *SCS National Bulletin* 300-2-7 (1981): 6, 8 (quote), https://www.nrcs.usda.gov/wps/portal/nrcs/detail/national/about/history/?&cid=nrcs143_021382, accessed August 2018.

6. Minutes, Colorado Association of Soil Conservation Districts, January 29, 1957, box 2, folder 16, 191, Records of the Colorado Association of Soil Conservation Districts, Water Resources Archive, Colorado State University, Fort Collins (hereafter Records of CASCD).

7. Minutes, CASCD, February 18, 1946, box 2, folder 7, 2–6, Records of CASCD.

8. Certificate of Incorporation, CASCD [1946], box 2, folder 5, Records of CASCD.

9. "Brief History of the Colorado Association of Soil Conservation Districts," box 2, folder 17, minute book, 297–98, Records of CASCD.

10. Stanley W. Voelker, "Land-Use Ordinances of Soil Conservation Districts in Colorado," *Colorado Agricultural Experiment Station Technical Bulletin* 45 (1952): 1–3; George C. Elliott, president's report, January 29, 1952, box 2, folder 17, 1, Records of CASCD.

11. Minutes, CASCD, February 18, 1946, box 2, folder 4, 6; State Conservation Board notes, 1947, box 2, folder 15; minutes, CASCD, January 4, 1954, box 2, folder 17, minute book, 192–94, all in Records of CASCD.

12. Lowell Watts, address to CASCD directors, July 14, 1959, box 2, folder 8, minute book, 3–4; George L. James, Weld County, "Annual Report [1946]," box 92, folder 8, 44–45, "Annual Report [1948]," box 93, folder 2, preface, and "Annual Report [1953]," box 94, folder 2, 32, all in Records of CES.

13. Minutes, CASCD, January 4, 1954, box 2, folder 17, minute book, 232; C. L. Terrill [extension conservationist] to CASCD board, undated, 1955, box 2, folder 17, minute book, 30, 38; president's annual report, February 1, 1956, box 2, folder 18, minute book 366, all in Records of CASCD.

14. Chester R. Fithian, Baca County, "Annual Report [1954]," box 9, folder 5, 51, Records of CES; Helms, "Great Plains Conservation Program," 6.

15. George C. Elliott, speech to CASCD, February 1, 1950, box 2, folder 3, 1–5 (quotes, p. 5), Records of CASCD.

16. George L. James, Weld County, "Annual Report [1952]," box 94, folder 1, 81, Records of CES; R. H. Foote, "The History of Artificial Insemination: Selected Notes and Notables," *Journal of the American Society of Animal Science* (2002): 1–2, 8, asas.org, accessed August 2021.

17. "Tenth Anniversary Colorado A&M Research on Improvement of Beef Cattle through Breeding," flyer in J. S Brinks, David W. Schafer, and Howard H. Stonaker, *Fifty Years of Bull Performance Tests and Research Studies at the San Juan Basin Research Center, Hesperus, Colorado*, vol. 1 (Fort Collins: Colorado State University, Animal Science Department, [1950?]).

18. Glen J. Gausman, W. H. Leonard, and Herman Fauber, "Test of Hybrid Corn under Irrigation in Colorado in 1947," Colorado Agricultural Experiment Station, *Miscellaneous Series* 390 (January 1948): 1.

19. Lana Reid, "How Corn Hybrids Are Developed" (Ottawa, ON: Research and Development Center, 2004), 1–3; Claudia Reinhardt and Bill Ganzel, "The Science of Hybrids" (York, NE: Living History Farm, 2003), 1–2, https://livinghistoryfarm.org/farminginthe30s/crops_03html, accessed August 2018.

20. Almond M. Binkley, "The Development of Onion Varieties and Hybrids Adapted to the Major Onion Growing Areas of Colorado [Project Outline, 1956]," box 80, folder 44, 2, Records of the Colorado Agricultural Experiment Station, Agricultural and Natural Resources Archive, Colorado State University, Fort Collins (hereafter Records of CAES); Charles F. Drage, Jefferson County, "Annual Report [1949]," box 121, folder 3, 14, Records of CES.

21. James F. Ellis, Mountain States Hardware and Implement Association, annual meeting, December 17, 1952, box 2, folder 17, 5, Records of CASCD.

22. Russell R. Poynor, "Agriculture, the Rediscovered Profession," annual meeting, CASCD, Denver, January 6, 1967, box 3, folder 5, minute book, 74 (first quote), 75 (second quote), 77 (third quote), 78–79 (fourth quote), Records of CASCD.

23. US Census 1954 and 1974.

24. A. F. Hoffman, "The Use of Commercial Fertilizers [1946]," box 20, folder 10, 28–31 (quote, p. 28), Records of CES.

25. *Colorado Agricultural Handbook* (Fort Collins: Colorado A&M, 1952), 6–7; George L. James, Weld County, "Annual Report [1947]," box 92, folder 10, 95, Records of CES.

26. W. Bentley Greb, "Significant Research Findings and Observations from the U.S. Central Great Plains Research Station and Colorado State University Experiment Station Cooperating, Akron, Colorado, Historical Summary, 1900–1981" (mimeograph) (Akron, CO: US Central Great Plains Research Station, 1981), 7.

27. James E. Hughs, Baca County, "Annual Report [1949]," box 8, folder 59, and James W. Reed, Cheyenne County, "Annual Report [1958]," box 16, folder 54, both in Records of CES; James E. Hansen III, *Beyond the Ivory Tower: A History of Colorado State University Cooperative Extension* (Fort Collins: Colorado State University, 1991), 67.

28. Bruce Jay Thornton, "2, 4-D Is Best Weed Control, Yet Has Definite Limitations," Colorado Agricultural Experiment Station, *Miscellaneous Series* 342 (October 1946): 1–2; R. H. Tucker, "Annual Report [1945]," box 111, folder 7, 121, Records of CES.

29. Edwin B. Colette, Boulder County, "Annual Report [1951]," box 13, folder 6, 31, Records of CES.

30. George L. James, Weld County, "Annual Report [1962]," box 95, folder 7, 2–3 (quote, p. 2), Records of CES; Legislative Committee minutes, January 9, 1959, box 2, folder 18, 4–9, Records of CASCD.

31. Convention flyer included in George L. James, Weld County, "Annual Report [1965]," box 96, folder 1, 482–83, Records of CES.

32. Robert L. Zimdahl, *Weeds of Colorado* (Fort Collins: Colorado State University Cooperative Extension Service, 1998); see also Zimdahl, *Agriculture's Ethical Horizon*, 2nd ed. (Amsterdam: Elsevier, 2012), 145.

33. Charles M. Drage, Jefferson County, "Annual Report [1948]," box 120, folder 2, 46, and Carl H. Powell, Delta County, "Annual Report [1955]," box 20, folder 20, 20, both in Records of CES.

34. Caitlin Coleman, *Citizen's Guide to Colorado's Transbasin Diversions* (Denver: Water Education Colorado, 2014), 4.

35. Raymond L. Anderson and Loyal M. Hartman, "Introduction of Supplemental Irrigation Water: Agricultural Response to an Increased Water Supply in Northeastern Colorado," *Colorado Agricultural Experiment Station Technical Bulletin* 76 (June 1965): 1–2.

36. Steven F. Mehls, *The New Empire of the Rockies: A History of Northeast Colorado* (Denver: US Bureau of Land Management, 1984), 174; Arwin M. Bolin, Baca County, "Annual Report [1957]," box 9, folder 4, 5–6, Records of CES.

37. Robert G. Evans, "Center Pivot Irrigation," draft, USDA Agricultural Research Service, Northern Plains Agricultural Research Laboratory, 2001, 1, 4, https://www.ars.usda.gov/ARSUserFiles/21563/center%20pivot%20design%202.pdf, accessed August 2018; Shelli Mader, "Center Pivot Irrigation Revolutionized Agriculture," *Fence Post* (Greeley, CO), May 25, 2010, https://www.thefencepost.com/news/center-pivot-irrigation-revolutionizes-agriculture/, accessed August 2018.

38. William E. Code, "Water Table Fluctuations in Eastern Colorado," *Colorado Agricultural Experiment Station Bulletin* 500-S (August 1958): 1–2; Morton W. Bittinger, "Colorado's Ground-Water Problems. Ground Water in Colorado," *Colorado Agricultural Experiment Station Bulletin* 504-S (1959): 17.

39. P. Andrew Jones and Tom Cech, *Colorado Water Law for Non-Lawyers* (Boulder: University Press of Colorado, 2009), 85, 133, 150–51, 154; Gregory J. Hobbs, "An Overview of Colorado Groundwater Law," *Colorado Ground-Water Association Newsletter* (Summer 2007): 1, 4–5, 9–10, 12.

40. John F. Freeman, *Persistent Progressives: The Rocky Mountain Farmers Union* (Boulder: University Press of Colorado, 2016), 118.

41. Freeman, *Persistent Progressives*, 119–20 (quote, p. 120).

42. Thomas E. Howard, *Agricultural Handbook for Rural Pastors and Laymen: Religious, Economic, Social, and Cultural Implications of Rural Life* (Des Moines, IA: National Catholic Rural Life Conference, 1946), 4 (first quote), 64 (second quote).

43. Howard, *Agricultural Handbook*, 34, 63 (quote).

44. Thomas J. Noel, *Colorado Catholicism and the Archdiocese of Denver, 1857–1989* (Boulder: University Press of Colorado, 1989), 144–45.

45. Corby Kummer, "Back to Grass," *Atlantic Monthly* 291 (May 2003): 138–42; Eric Schlosser, *Fast Food Nation: The Dark Side of the All-American Meal* (New York: Houghton Mifflin Harcourt, 2012 [2001]), 255.

46. Thomas M. Lasater, address to annual meeting, CASCD, Denver, January 4, 1968, box 3, folder 6, 76–77, Records of CASCD.

47. Lasater, address, 78, 80; see Ronald S. Knollenberg, "Environmental-Technological Interactions in Colorado High Plains Agricultural History," PhD dissertation, University of Colorado, Boulder, 1996, 159.

48. Steven F. Mehls, *The Valley of Opportunity: A History of West-Central Colorado* (Denver: US Bureau of Land Management, 1982), 264–67.

49. Mehls, *Valley of Opportunity*, 268.

50. Charles M. Nelson, Boulder County, "Annual Report [1957]," box 13, folder 11, 1, and Charles Bliss(?), Boulder County, "1973–74 Plan of Work," box 13, folder 26, 1, both in Records of CES; on Gundell, https://history.denverlibrary.org/news/almost-time-garden, accessed September 2018.

51. US Census 1950 and 1970; Raymond L. Anderson, "Urbanization of Rural Lands in the Northern Colorado Front Range," USDA Natural Resource Economic, Statistics, and Cooperative Service, Washington, DC, in conjunction with Colorado State University Cooperative Extension Service, Fort Collins, 1984, 12; Daniel Tyler, *The Last Water Hole in the West: The Colorado–Big Thompson Project and the Northern Colorado Water Conservancy District* (Niwot: University Press of Colorado, 1992), 267, 353.

52. C. Ivan Archer, Jefferson County, "Plan of Work [1981]," box 38, folder 24, 6, Records of CES; John Litz, Golden, CO, personal communication to John Freeman, September 2018.

53. Joseph B. Dischinger, "Local Government Regulations Using 1041 Powers," *Colorado Lawyer* 34, no. 12 (December 2005): 79–80; *Colorado Farm Bureau News*, March 1975, 6–7.

54. Harry R. Woodward Jr., address to annual meeting, January 7, 1969, box 3, folder 7, 24, Records of CASCD.

55. Quoted in Nell Brown Propst, *Grassroots People* (Denver: Colorado Farm Bureau, 1997), 200–201.

56. Robert Evans, "Conservation, the Foundation for Your Future [1965]," box 3, folder 3, 71–72, Records of CASCD.

57. Rev. Daniel O. Parker, "Soil Stewardship in an Urban Age," address to annual meeting, January 8, 1969, box 3, folder 7, 87, Records of CASCD.

58. Parker, "Soil Stewardship," 88–89.

Chapter 6: The Specter of Nature's Limits

1. Richard D. Lamm, "Goals and Objectives for Colorado's Long-Range Growth and Development," Denver, governor's office, September 10, 1976; see Lamm, "Forward," in Hershel Elliott, *Ethics for a Finite World: An Essay Concerning a Sustainable Future* (Golden, CO: Fulcrum, 2005), ix–xxiii.

2. Richard D. Lamm, Denver, discussion with John Freeman, June 2018.

3. Advertisement in the *Denver Post*, April 22, 1971, 10; "The Lamm Administration: A Retrospective," Governor's Office, Denver, 1986, 133–34; John N. Stencel III, discussion with John Freeman, June 2018.

4. "Governor Lamm Speaks on Land, Water and Other Things," *Colorado Farm Bureau News* (May 1976): 4.

5. "Governor Lamm Speaks," 13.

6. Lamm, discussion with Freeman, June 2018; news release on Governor Lamm's address to annual convention, November 29, 1976, box 7, folder 9, Records

of the Rocky Mountain Farmers Union, Agricultural and Natural Resources Archives, Archives and Special Collections, Colorado State University, Fort Collins.

7. US Census 1970 and 2000; see *Losing Ground: Colorado's Vanishing Agricultural Landscape* (Denver: Environment Colorado Research and Policy Center, 2006), 1.

8. John F. Freeman, *Persistent Progressives: The Rocky Mountain Farmers Union* (Boulder: University Press of Colorado, 2016), 163–65.

9. Virginia Culver, [John Fetcher obituary], *Denver Post*, February 10, 2009, https://www.denverpost.com/2009/02/10/engineer-found-his-home-on-steamboat-ranch/, accessed December 2018.

10. Merry Davis, "County Perspectives: A Report on 35 Acre Subdivision Exemptions in Colorado" (Denver: Colorado Counties, Inc., 2006), 2–3, https://www.colorado.gov/pacific/sites/default/files/13WaterResources1009CCI%20Report%20County%20Perspectives%20on%2035%20acre%20subdivision%20exemption%20in%20CO.pdf, accessed December 2018; Hal Clifford, "Saving the Ranch," *High Country News*, November 27, 1995, https://www.hcn.org/issues/48/1468, accessed December 2018.

11. Clifford, "Saving the Ranch."

12. Clifford, "Saving the Ranch."

13. C. J. Mucklow, discussion with John Freeman, June 2018; "A Guide to Rural Living and Small-Scale Agriculture," http://routt.extension.colostate.edu/agriculture/rural-living/, accessed June 2018; Hal Clifford, "Rancher's New Cash Crop Will Be Scenery," *High Country News*, November 27, 1995, https://www.hcn.org/issues/48, accessed December 2018.

14. Marsha Daughenbaugh and Todd Hagenbuch (county agent), discussion with John Freeman, October 2015; https://communityagalliance.org/, accessed October 2015.

15. Daughenbaugh and Hagenbuch, discussion with Freeman, October 2015.

16. Freeman, *Persistent Progressives*, 138–39, 166–67.

17. Paul K. Conkin, *A Revolution Down on the Farm: The Transformation of American Agriculture since 1929* (Louisville: University of Kentucky Press, 2008), 131.

18. Glen R. "Spike" Ausmus, discussion with John Freeman, October 2018; Freeman, *Persistent Progressives*, 126–27, 169.

19. Charles A. Francis, Raymond P. Poincelot, and George W. Bird, eds., *Developing and Extending Sustainable Agriculture, a New Social Contract* (New York: Haworth Food and Agricultural Products Press, 2006), xvii; James F. Parr, "USDA Research on Organic Farming: Better Late than Never," *American Journal of Alternative Agriculture* 18, no. 3 (2003): 171–72.

20. Vaclav Smil, *Enriching the Earth: Fritz Haber, Carl Bosch, and the Transformation of World Food Production* (Cambridge, MA: MIT Press, 2001), 166, 201–6, 217.

21. Abdelfettah Berrada, principal investigator, "Developing Sustainable Dryland Cropping Systems in Southwest Colorado and Southeast Utah Using Conservation Tillage and Crop Diversification," *Colorado Agricultural Experiment Station Technical Bulletin* TB02–02 (May 2002): 1–2, 26–27.

22. Virgil Kochis, discussion with John Freeman, October 2018; Bentley W. Greb, "Reducing Drought Effects on Croplands in the West-Central Great Plains," *USDA Agricultural Information Bulletin* 420 (June 1979): 1.

23. Quoted in William E. Pabor, *Colorado as an Agricultural State: Its Farms, Fields, and Garden Lands* (New York: Orange Judd, 1883), 12.

24. Kochis, discussion with Freeman, October 2018.

25. Freeman, *Persistent Progressives*, 42; Brad Erker (executive director, Colorado Wheat), "COAxium™ Wheat Production System," slide presentation to Independent Agricultural Consultants of Colorado, Denver, January 8, 2018, 7–8.

26. Erker, "COAxium™ Wheat Production System," 8–9.

27. Brad Erker, discussion with John Freeman, September 2018; Erker, "COAxium™ Wheat Production System," 9–11.

28. Erker, discussion with John Freeman, September 2018.

29. Kochis, discussion with Freeman, October 2018.

30. Kochis, discussion with Freeman, October 2018.

31. Jan Kochis to the *Colorado Springs Gazette*, June 30, 2018.

32. Ausmus, discussion with Freeman, October 2018; Ronald S. Knollenberg, "Environmental-Technological Interactions in Colorado High Plains Agricultural History," PhD dissertation, University of Colorado, Boulder, 1996, 438–39.

33. James Valliant, "Management Practices for Drip Irrigation in Baca County, Colorado," *Colorado Agricultural Experiment Station Technical Report* TR-07–13 (June 2007): 1–2.

34. Valliant, "Management Practices for Drip Irrigation," 4, 12.

35. Deborah E. Popper and Frank Popper, "The Great Plains: From Dust to Dust," *Planning Magazine* 53, no. 12 (December 1987): 12–18; https://landinstitute.org/our-work/perennial-crops/kernza/, accessed January 2019.

36. Matthew D. Heimerich, discussion with John Freeman, October 2018.

37. Freeman, *Persistent Progressives*, 131.

38. Freeman, *Persistent Progressives*, 132–33 (quote, p. 132); *City of Thornton v. Farmers Reservoir and Irrigation Company, "Judgment to Vacate Order of District Court," of Colorado*, 575 P.2nd 382 (1978), March 6, 1978, http://www.leagle.com/decision/1978957575P2d382_1948, accessed January 2019.

39. John F. Freeman, *High Plains Horticulture: A History* (Boulder: University Press of Colorado, 2008), 228–31.

40. Daniel Tyler, *The Last Water Hole in the West: The Colorado–Big Thompson Project and the Northern Colorado Water Conservancy District* (Niwot: University Press of Colorado, 1992), 447, 449–51; Heimerich, discussion with Freeman, October 2018.

41. Pastor Travis Walker and Kristina Walker, discussion with John Freeman, October 2018.

42. Caitlin Coleman, *Citizen's Guide to Colorado's Transbasin Diversion* (Denver: Water Education Colorado, 2014), 3, 18; Jay and De Bond, "Crowley County History," www.coloradoplains.com/crowley/history.htm, accessed January 2018.

43. Quoted in Marianne Goodland, "Buying and Drying: Water Lessons from Crowley County," https://www.coloradoindependent.com/2015/07/09/buying-and-drying-water-lessons-from-crowley-county/, accessed October 2018.

44. Quoted in Jim Robbins, "Water Sales Drain Life from Rural Colorado County," *Chicago Tribune*, August 27, 1992, https://www.chicagotribune.com/news/ct-xpm-1992-08-27-9203180177-story.html, accessed January 2018; Heimerich, discussion with Freeman, October 2018.

45. P. Andrew Jones and Tom Cech, *Colorado Water Law for Non-Lawyers* (Boulder: University Press of Colorado, 2009), 204, 222–23; Matthew D. Heimerich quoted in *Colorado Politics*, September 30, 2018, https://coloradopolitics.com/nyquist-colorado-springs-utilities-water/, accessed October 2018.

46. Heimerich, discussion with Freeman, October 2018.

47. Scott Campbell, Kevin League, and Nathan Meyer, "Our Land, Our Water, Our Future: A Conservation Plan for the Western Lower Arkansas Valley" (Colorado Springs: Palmer Land Trust, 2012).

48. Heimerich, discussion with Freeman, October 2018.

49. Travis and Kristina Walker, discussion with Freeman, October 2018.

50. Robert R. Crifasi, *A Land Made from Water: Appropriation and the Evolution of Colorado's Landscape, Ditches, and Water Institutions* (Boulder: University Press of Colorado, 2015), 311–13 (quotes, p. 311).

51. Michelle Bryan, "Agricultural Mitigation Case Studies: Program Summaries and Stakeholder Perspectives from Seven Western Communities" (Missoula: University of Montana School of Law, 2015), 14–19, https://farmlandinfo.org/publications/agricultural-mitigation-case-studies-program-summaries-and-stakeholder-perspectives-from-seven-western-communities/-usit, accessed January 2019; section on Boulder County in "Practical Guide to Transfer of Development Credits (TDCs) in Alberta" (Calgary, Alberta: Miistakis Institute, 2013), http://www.tdc-alberta.ca/pieces_egs_boulder.html, accessed January 2019; see Brian Halweil and Danielle Nierenberg, "Farming the Cities," in Linda Starke, ed., *2007 State of the World: Our Urban Future, Worldwatch Institute Report on Progress Toward a Sustainable Society* (New York: Norton, 2007), 48–50.

52. Deni LaRue, "Right to Farm Summary," prepared for the Board of the Larimer County Commissioners, August 1998, https://www.larimer.org/policies/right-to-farm-summary, accessed January 2019.

53. Philip Conford, *The Origins of the Organic Movement* (Edinburgh: Floris, 2001), 17–19.

54. Conford, *Origins of the Organic Movement*, 99–101; Andrew N. Case, *Rodale and the Making of Marketplace Environmentalism* (Seattle: University of Washington Press, 2018), 208.

55. USDA Study Team on Organic Farming, "Report and Recommendations on Organic Farming," July 1980, xi–xii (quote, p. xi), usda_organic_1980.pdf (centerforinquiry.org), accessed October 2018, August 2021.

56. John Ellis, discussion with John Freeman, February 2016.

57. Freeman, *Persistent Progressives*, 150–51.

58. Freeman, *Persistent Progressives*, 156–62.

Chapter 7: Organic by Choice

1. Reverend Stephanie L. Price, discussion with John Freeman and Mark Uchanski, July 2018; "The Land: A United Methodist Faith Community," 1–4, https://www.hopechangeslives.org/wp-content/uploads/2015/10/Land_Implementation-Grant.pdf, accessed June 2019.

2. Stephanie Price, "Planting Seeds in 2019," 1, https://thelandaurora.org/planting-seeds-in-2019/, accessed June 2019.

3. Dave Carter, discussion with Mark Uchanski, May 2019; Organic Foods Production Act, Title 21, Food, Agriculture, Conservation, and Trade Act of 1990, codified at 7 U.S.C. §7 U.S.C. ch. 94, 6501 et seq., https://www.law.cornell.edu/uscode/text/7/chapter-94, accessed June 2019.

4. USDA-SARE, "Opportunities in Agriculture: Transitioning to Organic Production," 2012, 2, https://www.sare.org/Learning-Center/Bulletins/Transitioning-to-Organic-Production, accessed April 2019.

5. John P. Reganold and Jonathan M. Wachter, "Organic Agriculture in the Twenty-First Century," *Nature Plants* 2 (February 2016): 1, https://www.researchgate.net/publication/293014068_Organic_agriculture_in_the_twenty-first_century, accessed April 2019.

6. "Farms, Land Use, and Sales of Organically Produced Commodities on Certified and Exempt Organic Farms: 2008," 1–9, https://www.nass.usda.gov/Publications/AgCensus/2007/Online_Highlights/Organics/organics_1_01.pdf; "Certified Organic Survey 2016 Summary" v–vii, https://www.nass.usda.gov/Publi

cations/Todays_Reports/reports/census17.pdf.; "Colorado Organic Industry Overview (2013?)," https://www.colorado.gov/pacific/sites/default/files/Colorado%20Organic%20Industry%20Overview.pdf, all accessed June 2019.

7. Dell Rae Moellenberg, "Lewis O. Grant Obituary," August 2, 2013, news release, Office of the Provost, Colorado State University, Fort Collins.

8. "A Visit to Grant Family Farms in Colorado," October 19, 2012, https://www.farmaid.org/blog/a-visit-to-grant-family-farms-in-colorado/; David Migoya, "Bankrupt Grant Family Farms Had Owed Nearly 10 Million," https://www.denverpost.com/2013/01/16/bankrupt-grant-family-farms-had-owed-nearly-10-million/, both accessed June 2019.

9. Ferrell quoted in Max Siegelbaum, "Colorado Farmers Going Organic to Meet Rising Demand," *Denver Post*, February 2, 2017, https://www.denverpost.com/2017/02/07/colorado-farmers-going-organic/; https://www.berrypatchfarms.com/our-story-1; https://monroefarm.com/our-story, all accessed June 2019.

10. https://www.hungenbergproduce.com/history/; Stephanie Strong, "Business Paying Farmers to Go Organic Even Before the Crops Come In," https://www.nytimes.com/2016/07/15/business/paying-farmers-to-go-organic-even-before-the-crops-come-in.html, accessed June 2019.

11. Jean Hediger, discussions with Mark Uchanski and students, March 2018.

12. https://www.businesswire.com/news/home/20110922005517/en/Horizon%C2%AE-Celebrates-20-Years-Organic-Leadership; White Wave Historical Information | Danone North America, both accessed August 2021.

13. "Mark A. Retzloff," https://www.bloomberg.com/research/stocks/private/person.asp?personId=386199&privcapId=114919867.

14. "Aurora Organic Dairy," https://www.auroraorganic.com, accessed June 2019.

15. "Naturally Boulder," https://www.naturallyboulder.org, accessed June 2019.

16. John F. Freeman, *Persistent Progressives: The Rocky Mountain Farmers Union* (Boulder: University Press of Colorado, 2016), 215–16.

17. Press release, Naturally Boulder, August 30, 2011, https://www.naturallyboulder.org/press-release/gmo-crops-on-boulder-county-open-space/; Sam Lounsberry, "Boulder County Commissioners Set to Consider Proposal Delaying Open Space Phaseout of GMOs," *Longmont Times-Call*, May 7, 2019, https://www.timescall.com/2019/05/07/boulder-county-commissioners-set-to-consider-proposal-delaying-open-space-phase-out-of-genetically-modified-crops/, accessed June 2019.

18. Robert L. Zimdahl, *Agriculture's Ethical Horizon*, 2nd ed. (Amsterdam: Elsevier, 2012), 154.

19. Theodore M. Klein and Todd J. Jones, "Methods of Genetic Transformation: The Gene Gun," in Indra K. Vasil, ed., *Molecular Improvement of Cereal Crops* (Dordrecht: The Netherlands, 1999), 25, 35.

20. H. Daniel, B. Muthukumar, and S. B. Lee, "Marker Free Transgenic Plants: Engineering the Chloroplast Genome without the Use of Antibiotic Selection," *Current Genetics* 39 (2001): 109, SpringerLink | Code of Federal Regulations, Part 2015 National Organic Program "Excluded Methods": 3CFR : 7 CFT Part 205—National Organic Program (federalregister.gov), both accessed June 2019.

21. Andrew Hopp, discussion with Mark Uchanski, May 2019; Stanton B. Gelvin, "*Agrobacterium*-Mediated Plant Transformation: The Biology behind the 'Gene-Jockeying' Tool," *Microbiology and Molecular Biology Reviews* 67 (2003): 16–20, https://www.ncbi.nlm.nih.gov/pmc/articles/PMC150518/, accessed August 2021.

22. Hopp, discussion with Uchanski, May 2019; Jesus Madrazo, "An Open Letter to Our Stakeholders," also run as an advertisement in the *New York Times*, October 31, 2018, https://www.bayer.com/en/our-vision-for-safe-and-sustainable-agriculture, accessed June 2019.

23. Wenyuan Han and Qunxin She, "Chapter One—CRISPR History: Discovery, Characterization, and Prosperity," *Progress in Molecular Biology and Translational Science* 152 (2017): 2–4, https://ebin.pub/crispr-in-animals-and-animal-models-volume-152-0128125063-9780128125069.html, accessed May 2019.

24. "Biotech Firm Calyxt Launches Soybean Oil Made from Gene-Edited Soybeans," https://geneticliteracyproject.org/2019/03/01/biotech-firm-calyxt-launches-soybean-oil-made-from-gene-edited-soybeans/, accessed June 2019.

25. Specialty crop definition from 2004 Farm Bill, https://www.ams.usda.gov/sites/default/files/media/USDASpecialtyCropDefinition.pdf, accessed June 2019.

26. Frank Stonaker, discussion with John Freeman, October 2015.

27. Stonaker, discussion with Freeman, October 2015.

28. Stonaker, discussion with Freeman, October 2015; Frank Stonaker, "A Guide for Small-Scale Organic Vegetable Farmers in the Rocky Mountain Region," PhD dissertation, Colorado State University, Fort Collins, 2009; Myron F. Babb and James E. Kraus, "Home Vegetable Gardening in the Central and High Plains and Mountain Valleys," *USDA Farmers' Bulletin* 2000 (1949), 1–98.

29. Dan Hobbs, discussion with John Freeman, May 2013.

30. Hobbs, discussions with Freeman and Uchanski, October 2018.

31. Michael Bartolo, discussion with John Freeman and Mark Uchanski, October 2018.

32. Luke Runyan, "Cantaloupe Growers Plead Guilty to Criminal Charges," Harvest Public Media, May 22, 2013, https://www.kunc.org/health/2013-5-22/cantaloupe-farmers-plead-guilty-to-criminal-charges, accessed October 2018.

33. USDA ERS, "Agricultural Act of 2014: Highlights and Implications," USDA ERS-Agricultural Act of 2014: Highlights and Implications, https://www.ers.usda.gov/agricultural-act-of-2014-highlights-and-implications/; Jennifer Steinhauer,

"Farm Bill Reflects Shifting American Menu and a Senator's Persistent Tilling," *New York Times*, March 9, 2014, https://www.nytimes.com/2014/03/09/us/politics/farm-bill-reflects-shifting-american-menu-and-a-senators-persistent-tilling.html, both accessed June 2019.

34. "Statement of Principles on Industrial Hemp," https://www/federalregister.gov/documents/2016/08/12/2016-19146/statement-of-principles-on-industrial-hemp, accessed June 2019.

35. Abdel F. Berada, ed., "2014–2018 Research Results, Southwestern Colorado Research Center," *Colorado Agricultural Experiment Station Technical Report* TR19–03 (2019): 5–7.

36. Jack Healy, "Colorado's Marijuana Experiment, after Five Years," *New York Times*, July 1, 2019, A1, A12–13.

37. https://www.nasda.org/person/kate-greenberg, accessed August 2021.

38. "Governor Polis Releases Roadmap to 100 Percent Renewable Energy and Bold Climate Action," Governor's Office, May 30, 2019, https://www.colorado.gov/governor/news/governor-polis-releases-roadmap-100-percent-renewable-energy-and-bold-climate-action, accessed April 2020; Sam Lounsberry, "Potential of Marriage between Solar Power, Agriculture, Beehives to Be Studied Just Outside Longmont," *Longmont Times-Call*, March 25, 2019, https://www.timescall.com/2019/03/25/potential-of-marriage-between-solar-power-agriculture-beehives-to-be-studied-just-outside-longmont/, accessed April 2020.

39. Courtney White, "Agrovoltaics," *Conservation Magazine* (University of Washington) (July 2014), https://www.conservationmagazine.org/2014/07/agrivoltaics/, accessed June 2019.

40. White, "Agrovoltaics"; "Jack's Solar Garden," https://www.jackssolargarden.com, accessed June 2019.

41. https://sandboxsolar.com/colorado-solar-research-csu/, accessed August 2021.

42. Farm Shares, "Browse CSAS in Colorado by Pickup City," https://farmshares/browse-CSAs-in-Colorado-by-pickup-city/; USDA, "Farm to School Census," Colorado | USDA-FNS Farm to School Census, https://farmtoschoolcensus.fns.usda.gov, both accessed June 2019.

43. Blake Angelo, "Developing Denver's Food System Infrastructure," January 6, 2016, New & Next: Developing Denver's Food System Infrastructure (https://www.confluence-denver.com/featuredposts/new_and_next_angelo_010616.aspx); "Denver Food Action Plan," June 2018, DenverFoodActionPlan.pdf (https://www.denvergov.org/content/dam/denvergov/Portals/771/documents/CH/Food%20Action%20Plan/DenverFoodActionPlan.pdf), both accessed June 2019.

44. Gregory Graff, Ryan Mortenson, Rebecca Goldbach, Dawn Thilmany, Stephen Davies, Stephen Koontz, Geniphyr Ponce-Pore, and Kathay Rennels, *The Value Chain of Colorado Agriculture* (Fort Collins: CSU Department of Agricultural and Resource Economics, 2013), iii–viii; Blake Angelo, "Colorado Is an Agricultural State," March 12, 2016, Blake Angelo Current Thoughts—Blake Angelo, http://www.blakeangelo.com/thoughts/colorado-is-an-agricultural-state, accessed August 2021.

45. Vaclav Smil, *Enriching the Earth: Fritz Haber, Carl Bosch, and the Transformation of World Food Production* (Cambridge, MA: MIT Press, 2001), 164–65.

46. National Research Council, *Toward Sustainable Agricultural Systems in the 21st Century* (Washington, DC: National Academies Press, 2010), 4; John P. Reganold and Jonathan M. Wachter, "Organic Agriculture in the Twenty-First Century," *Nature Plants* 2 (February 2016): 1.

47. https://www.nass.usda.gov/Publications/AgCensus/2017/Online_Resources/Organics/ORGANICS.txt; https://www.revision.coop/; https://www.wedontwaste.org/, all accessed August 2021.

48. Larry Vickerman, "Regenerative Agriculture," *Inside the Gardens* (Summer 2019): 4.

49. Reganold and Wachter, "Organic Agriculture," 2–5.

50. Richard D. Lamm, "A New Moral Vision for the Finite World," unpublished manuscript, 23, used with author's permission (first quote); foreword to Herschel Elliott, *Ethics for a Finite World: An Essay Concerning a Sustainable Future* (Golden, CO: Fulcrum, 2005), xxii–iii (second quote).

51. "Governor Lamm Speaks on Land, Water, and Other Things," *Colorado Farm Bureau News*, May 1976, 13.

Epilogue

1. James E. Payne, "Field Notes from Trips in Eastern Colorado," *Colorado Agricultural Experiment Station Bulletin* 59 (December 1900): 7–9.

2. "Colorado Land Link," https://guidestonecolorado.org/colorado-land-link/, accessed April 2020; see John F. Freeman, *Persistent Progressives: The Rocky Mountain Farmers Union* (Boulder: University Press of Colorado, 2016), 205.

3. Richard D. Lamm, "A New Moral Vision for a Finite World," unpublished manuscript (2002?), 23.

4. "Definition," *Research and Degrowth*, Definition—Research and Degrowth (R&D), accessed August 2021.

Bibliography

The principal sources for this book are the collections of county agent and experiment station papers and reports, part of the Agricultural and Natural Resources Archive, Archives and Special Collections Department, Morgan Library, Colorado State University, Fort Collins, supplemented by US Census data and interviews with agriculturists throughout Colorado. Many of the historical collections are now available online, starting with https://mountainscholar.org/handle/10217/100005.

A word on federal records: until 1950, the US Census Bureau took the census of agriculture every ten years; between 1954 and 1975, it did so only in the years ending in 4 and 9 and in 1978, 1982, 1987, and 1992. The US Department of Agriculture's National Agriculture Statistics Service took over the census of agriculture from the Census Bureau beginning with the 1997 agriculture census. The 2017 census of agriculture was published in May 2019. Agricultural statistics for all census years are readily accessible through https://www.nass.usda.gov/AgCensus/.

https://doi.org/10.5876/9781646422050.c009

For the years prior to 1997, that website links to Census Bureau reports that are part of the Census of Agriculture Historical Archive, Albert R. Mann Library, Cornell University, Ithaca, NY. *U.S. Statutes at Large* can now be accessed at http://www.constitution.org/uslaw/sal/sal.htm.

Abbott, Carl, Stephen J. Leonard, and Thomas J. Noel. *Colorado: A History of the Centennial State*. 5th ed. Boulder: University Press of Colorado, 2013.

Albertson, Dean. *Roosevelt's Farmer: Claude R. Wickard in the New Deal*. New York: Columbia University Press, 1961.

Anderson, Raymond L. "Urbanization of Rural Lands in the Northern Colorado Front Range." Washington, DC: USDA Natural Resource Economics Division, Economics, Statistics, and Cooperative Service, in conjunction with Colorado State University Cooperative Extension Service, 1984.

Anderson, Raymond L., and Loyal M. Hartman. "Introduction of Supplemental Irrigation Water: Agricultural Response to an Increased Water Supply in Northeastern Colorado." *Colorado Agricultural Experiment Station Technical Bulletin* 76 (June 1965): 1–34.

Athearn, Frederic J. *Land of Contrast: A History of Southeast Colorado*. Denver: US Bureau of Land Management, 1985.

Babb, Myron F., and James E. Kraus. "Home Vegetable Gardening in the Central and High Plains and Mountain Valleys." *USDA Farmers' Bulletin* 2000 (1949): 1–98.

Bailey, Liberty Hyde. *The Holy Earth: Toward a New Environmental Ethic*. Introduction by Norman Wirzba. Mineola, NY: Dover, 2009.

Bailey, Liberty Hyde, ed. *Report of the Commission on Country Life*. New York: Sturgis and Walton, 1911.

Baker, Gladys L., Wayne D. Rasmussen, Vivian Wiser, and Jane M. Porter. *Century of Service: The First 100 Years of the United States Department of Agriculture*. Washington, DC: US Department of Agriculture, 1963.

Baumunk, Lowell, "Nathan Cook Meeker, Colonist." Master's thesis, Colorado State College of Education, Greeley, 1949.

Bennett, Hugh H., and William R. Chapline. "Soil Erosion: A National Menace." *USDA Circular* 33 (1928): 1–36.

Berman, David R. *Radicalism in the Mountain West, 1890–1920: Socialists, Populists, Miners, and Wobblies*. Boulder: University Press of Colorado, 2007.

Bickel, George L. *Rocky Mountain Farmers Union: A History 1907–1978*. Denver: Rocky Mountain Farmers Union, 1978.

Birkenfield, Daryl. "A Region Reforming: The Philosophy, Definition, and Brief History of Ogallala Commons." http://ogallalacommons.org/wp-content/uploads/2013/01/ARegionReforming-2.pdf. Accessed February 2019.

Bittinger, Morton W. "Colorado's Ground-Water Problems: Ground Water in Colorado." *Colorado Agricultural Experiment Station Bulletin* 504-S (1959): 1–28.

Blinn, Philo K. "Development of the Rocky Ford Cantaloupe Industry." *Colorado Agricultural Experiment Station Bulletin* 108 (March 1906): 1–17.

Blinn, Philo K. "A Rust-Resistant Cantaloupe." *Colorado Agricultural Experiment Station Bulletin* 104 (November 1905): 1–15.

Block, William Joseph. *The Separation of the Farm Bureau and the Extension Service: Political Issue in a Federal System*. Urbana: University of Illinois Press, 1960.

Boyd, David. *A History: Greeley and the Union Colony of Colorado*. Greeley: Greeley Tribune Press, 1890.

Brandon, Joseph F., and Alvin Kezer. "Soil Blowing and Its Control in Colorado." *Colorado Agricultural Experiment Station Bulletin* 419 (January 1936): 1–20.

Burdick, Raymond T. "Economics of Sugar Beet Production in Colorado." *Colorado Agricultural Experiment Station Bulletin* 453 (June 1939): 1–63.

Campbell, Scott, Kevin League, and Nathan Meyer. "Our Land, Our Water, Our Future: A Conservation Plan for the Western Lower Arkansas Valley." Colorado Springs: Palmer Land Trust, [2012]. https://www.palmerlandtrust.org/sites/default/files/media/PLTCONSERVPLANSPREADlr.pdf. Accessed January 2018.

Carter, David E. "Hard Choices: The Birth and Death of Ranchers' Choice Cooperative." In *The New Generation Cooperatives: Case Studies Expanded*, ed. Mary Holmes, Norman Walzer, and Christopher D. Merrit, 121–31. Macomb: Western Illinois University, 2001.

Case, Andrew. *Rodale and the Making of Marketplace Environmentalism*. Seattle: University of Washington Press, 2018.

Christenson, Reo Millard. *The Brannan Plan*. Ann Arbor: University of Michigan Press, 1959.

Code, William E. "Water Table Fluctuations in Eastern Colorado." *Colorado Agricultural Experiment Station Bulletin* 500-S (August 1958): 1–34.

Coleman, Caitlin, *Citizen's Guide to Colorado's Transbasin Diversion*. Denver: Water Education Colorado, 2014.

Conkin, Paul K. *A Revolution Down on the Farm: The Transformation of American Agriculture since 1929*. Louisville: University of Kentucky Press, 2008.

Cordell, Linda S., and Maxine E. McBrinn. *Archeology of the Southwest*. 3rd ed. Walnut Creek, CA: Left Coast Press, 2012.

Cornford, Philip. *The Origins of the Organic Movement*. Edinburgh: Floris, 2001.

Crifasi, Robert R. *A Land Made from Water: Appropriation and the Evolution of Colorado's Landscape, Ditches, and Water Institutions*. Boulder: University Press of Colorado, 2015.

Danbom, David B. *Born in the Country: A History of Rural America*. 2nd ed. Baltimore: Johns Hopkins University Press, 2006.

Davis, Ellen F. *Scripture, Culture, and Agriculture: An Agrarian Reading of the Bible*. Cambridge: Cambridge University Press, 2009.

Dean, Virgil W. "Charles F. Brannan and the Rise and Fall of Truman's 'Fair Deal' for Farmers." *Agricultural History* 69, no. 1 (Winter 1995): 28–53.

Dean, Virgil W. *An Opportunity Lost: The Truman Administration and the Farm Policy Debate*. Columbia: University of Missouri Press, 2006.

Dischinger, Joseph B. "Local Government Relations Using 1041 Powers." *Colorado Lawyer* 34, no. 12 (December 2005): 79–87.

Dunbar, Robert G. *Forging New Rights in Western Water*. Lincoln: University of Nebraska Press, 1983.

Dunbar, Robert G. "History of Agriculture." In *Colorado and Its People*, ed. Leroy Hafen, 2: 121–57. New York: Lewis Historical Publishing, 1948.

Dyson, Lowell K. *Farmers' Organizations*. New York: Greenwood, 1986.

Earp, Edwin L. *Biblical Backgrounds for the Rural Message*. New York: Association Press, 1922.

Elliott, Herschel. *Ethics for a Finite World: An Essay Concerning a Sustainable Future*. Golden, CO: Fulcrum, 2005.

Foote, R. H. "The History of Artificial Insemination: Selected Notes and Notables." *Journal of the American Society of Animal Science* (2002): 1–10. https://www.asas.org/docs/publications/footehist.pdf?sfvrsn=0. Accessed August 2018.

Freeman, John F. *High Plains Horticulture: A History*. Boulder: University Press of Colorado, 2008.

Freeman, John F. *Persistent Progressives: The Rocky Mountain Farmers Union*. Boulder: University Press of Colorado, 2016.

Graff, Gregory, Ryan Mortenson, Rebecca Goldbach, Dawn Thilmany, Stephen Davies, Stephen Koontz, Geniphyr Ponce-Pore, and Kathay Rennels. *The Value Chain of Colorado Agriculture*. Fort Collins: Colorado State University Department of Agricultural and Resource Economics, 2013. https://mountainscholar.org/bitstream/handle/10217/190028/FACFAGRE_ucsu522v232013.pdf?sequence=1&isAllowed=y. Accessed June 2019.

Greb, Bentley W. "Significant Research Findings and Observations from the U.S. Central Great Plains Research Station and Colorado State University Experiment Station Cooperating Akron, Colorado, Historical Summary, 1900–1981" (mimeograph). Akron, CO: US Central Great Plains Research Station, 1981.

Greeley, Horace. *An Overland Journey from New York to San Francisco in the Summer of 1859*, ed. Charles T. Duncan. New York: Alfred E. Knopf, 1964 [1860].

Greenbaum, Fred. *Fighting Progressive: A Biography of Edward P. Costigan*. Washington, DC: Public Affairs Press, 1971.

Hafen, Leroy R., ed. *Colorado and Its People*. 2 vols. New York: Lewis Historical Publishing, 1948.

Hansen, Herbert G. "Re-vegetation of Waste Range Land." *Colorado Agricultural Experiment Station Bulletin* 332 (January 1928): 1–10.

Hansen, James E., III. *Beyond the Ivory Tower: A History of Colorado State University Cooperative Extension*. Fort Collins: Colorado State University, 1991.

Hansen, James E., III. *Democracy's College in the Centennial State: A History of Colorado State University*. Fort Collins: Colorado State University, 1977.

Helms, Douglas. "The Preparation of the Standard State Soil Conservation Districts Law: An Interview with Philip M. Glick." Washington, DC: USDA Soil Conservation Service, February 1990. http://www.nrcs.usda.gov/Internet/FSE_DOCUMENTS/nrcs143_021270.pdf. Accessed October 2018.

Howard, Robert P. *James R. Howard and the Farm Bureau*. Ames: Iowa State University Press, 1983.

Howard, Thomas E. *Agricultural Handbook for Rural Pastors and Laymen: Religious, Economic, Social, and Cultural Implications of Rural Life*. Des Moines, IA: National Catholic Rural Life Conference, 1946.

Jack, Zachary Michael, ed. *Liberty Hyde Bailey: Essential Agrarian and Environmental Writings*. Ithaca, NY: Cornell University Press, 2008.

Jackson, Wes. *New Roots for Agriculture*. Lincoln: University of Nebraska Press, 1980.

Jones, Charles R. "Grasshopper Control." *Colorado Agricultural Experiment Station Bulletin* 233 (June 1917): 1–29.

Jones, P. Andrew, and Tom Cech. *Colorado Water Law for Non-Lawyers*. Boulder: University Press of Colorado, 2009.

Kindscher, Kelly. *Edible Wild Plants of the Prairie: An Ethnobotanical Guide*. Lawrence: University Press of Kansas, 1987.

Kluger, James R. *Turning on Water with a Shovel: The Career of Elwood Mead*. Albuquerque: University of New Mexico Press, 1992.

Lamm, Richard D., and Duane A. Smith. *Pioneers and Politicians: 10 Colorado Governors in Profile*. Boulder: Pruett, 1984.

Lansing, Michael J. *Insurgent Democracy: The Nonpartisan League in North American Politics*. Chicago: University of Chicago Press, 2015.

Lowe, Kevin M. *Baptized with the Soil: Christian Agrarians and the Crusade for Rural America*. New York: Oxford, 2016.

Lowitt, Richard, and Maurine Beasley, eds. *One-Third of a Nation: Lorena Hickok Reports on the Great Depression*. Urbana: University of Illinois Press, 1981.

May, William J. *The Great Western Sugarlands: The History of the Great Western Sugar Company and the Economic Development of the Great Plains*. New York: Garland, 1989.

McCarthy, G. Michael. *Hour of Trial: The Conservation Conflict in Colorado and the West, 1891–1907*. Norman: University of Oklahoma Press, 1977.

McCune, Wesley. *The Farm Bloc*. Garden City, NY: Doubleday, 1943.

McMath, Robert C., Jr. *American Populism: A Social History, 1877–1898*. New York: Hill and Wang, 1993.

Mehls, Steven F. *The New Empire of the Rockies: A History of Northeast Colorado*. Denver: US Bureau of Land Management, 1984.

Mehls, Steven F. *The Valley of Opportunity: A History of West-Central Colorado*. Denver: US Bureau of Land Management, 1982.

Minnis, Paul E., ed. *People and Plants in Ancient Western North America*. Washington, DC: Smithsonian Books, 2004.

National Research Council. *Toward Sustainable Agricultural Systems in the 21st Century*. Washington, DC: National Academies Press, 2010.

Noel, Thomas J. *Colorado Catholicism and the Archdiocese of Denver, 1857–1989*. Boulder: University Press of Colorado, 1989.

O'Rourke, Paul M. *Frontier in Transition: A History of Southwestern Colorado*. Denver: US Bureau of Land Management, 1980.

Pabor, William E. *Colorado as an Agricultural State: Its Farms, Fields, and Garden Lands*. New York: Orange Judd, 1883.

Pabor, William E. *First Annual Report of the Union Colony of Colorado*. New York: George W. Southwick, 1871.

Parshall, Ralph L. "The Parshall Measuring Flume." *Colorado Agricultural Experiment Station Bulletin* 423 (March 1936): 1–84.

Payne, James E. "Cattle Raising on the Plains." *Colorado Agricultural Experiment Station Bulletin* 87 (June 1904): 6–17.

Payne, James E. "Field Notes from Trips in Eastern Colorado." *Colorado Agricultural Experiment Station Bulletin* 59 (December 1900): 1–16.

Payne, James E. "Unirrigated Lands of Eastern Colorado." *Colorado Agricultural Experiment Station Bulletin* 77 (February 1903): 1–16.

Payne, James E. "Wheat Raising on the Plains." *Colorado Agricultural Experiment Station Bulletin* 89 (June 1904): 25–30.

Porter, Kimberly K. "Embracing the Pluralistic Perspective: The Iowa Farm Bureau and the McNary-Haugen Movement." *Agricultural History* 72, no. 2 (Spring 2000): 381–92.

Propst, Nell Brown. *Grassroots People*. Denver: Colorado Farm Bureau, 1997.

Rasmussen, Wayne D. *Taking the University to the People: Seventy-five Years of Cooperative Extension*. Ames: Iowa State University Press, 1989.

Rasmussen, Wayne D., Gladys L. Baker, and James S. Ward. "A Short History of Agricultural Adjustment, 1933–75." *USDA Economic Research Service Agriculture Information Bulletin* 391 (March 1976): 1–21.

Reganold, John P., and Jonathan M. Wachter. "Organic Agriculture in the Twenty-First Century." *Nature Plants* 2 (February 2016): 1–8.

Robbins, Wilfred W., and Breeze Boyack. "The Identification and Control of Colorado Weeds." *Colorado Agricultural Experiment Station Bulletin* 251 (July 1919): 1–128.

Saloutos, Theodore. *The American Farmer and the New Deal*. Ames: Iowa State University Press, 1982.

Samson, R. Neil. *For Love of the Land: A History of the National Association of Conservation Districts*. League City, TX: National Association of Conservation Districts, 1985.

Sandsten, Emil P. "Potato Growing in Colorado." *Colorado Agricultural Experiment Station Bulletin* 314 (January 1927): 1–31.

Schulte, Steven C. *Wayne Aspinall and the Shaping of the American West*. Boulder: University Press of Colorado, 2002.

Sears, Paul B. *Deserts on the March*. Norman: University of Oklahoma Press, 1935.

Sheflin, Douglas. "The New Deal Personified: A. J. Hamman and the Cooperative Extension Service in Colorado." *Agricultural History* 90, no. 3 (Summer 2016): 356–78.

Shover, John L. *Cornbelt Rebellion: The Farmers' Holiday Association*. Urbana: University of Illinois Press, 1965.

Smil, Vaclav. *Enriching the Earth: Fritz Haber, Carl Bosch, and the Transformation of World Food Production*. Cambridge, MA: MIT Press, 2001.

Smith, Kimberly K. *Wendell Berry and the Agrarian Tradition: A Common Grace*. Lawrence: University of Kansas Press, 2003.

Steinel, Alvin T. *History of Agriculture in Colorado*. Fort Collins, CO: State Agricultural College, 1926.

Stonaker, Frank. "A Guide for Small-Scale Organic Vegetable Farmers in the Rocky Mountain Region." PhD dissertation, Colorado State University, Fort Collins, 2009.

Summers, Thomas H., and R. W. Schafer, eds. *An Agricultural Program for Northwest Colorado*. Fort Collins: Colorado Agricultural College Extension Service, 1928.

Thornton, Bruce J. "The Colorado Pure Seed Law and Use of the Seed Laboratory." *Colorado Agricultural Experiment Station Miscellaneous Series* 129 (1942?): 1–3.

Tyler, Daniel. *The Last Water Hole in the West: The Colorado–Big Thompson Project and the Northern Colorado Water Conservancy District*. Niwot: University Press of Colorado, 1992.

US Department of Agriculture. "A Time to Act: A Report of the USDA National Commission on Small Farms." *USDA Miscellaneous Publications* 1545 (January 1998). http://www.csrees.usda.gov/nea/ag_systems/pdfs/time_to_act_1998.pdf. Accessed December 2013.

Valliant, James. "Management Practices for Drip Irrigation in Baca County, Colorado." *Colorado Agricultural Experiment Station Technical Report* TR-07-13 (June 2007): 1–30.

Wickens, James. *Colorado in the Great Depression*. New York: Garland, 1979.

Willard, James F., ed. *The Union Colony at Greeley, Colorado, 1869–1871*. Boulder: University of Colorado Historical Collection, 1918.

Williams, C. Fred. "William M. Jardine and the Foundations for Republican Farm Policy, 1925–1929." *Agricultural History* 70, no. 2 (Spring 1996): 216–32.

Wright, James E. *The Politics of Populism: Dissent in Colorado*. New Haven, CT: Yale University Press, 1974.

Zimdahl, Robert L. *Agriculture's Ethical Horizon*. 2nd ed. Amsterdam: Elsevier, 2012.

Zimdahl, Robert L. *Weeds of Colorado*. Fort Collins: Colorado State University Cooperative Extension Service, 1998.

Index

About the Authors

JOHN F. FREEMAN is the founder and president emeritus of the Wyoming Community Foundation. He has a PhD in early modern European history from the University of Michigan and is the author of *High Plains Horticulture: A History* and *Persistent Progressives: The Rocky Mountain Farmers Union*.

MARK E. UCHANSKI is an associate professor of horticulture at Colorado State University. He has a PhD in horticulture from the University of Illinois at Urbana-Champaign. He teaches, provides outreach, and conducts research in vegetable cropping systems and organic agriculture.